AI-Powered Ecommerce

How Machine Learning Is Transforming Online Shopping

Ramgopal Prajapat

Apress®

AI-Powered Ecommerce: How Machine Learning Is Transforming Online Shopping

Ramgopal Prajapat
Bangalore, Karnataka, India

ISBN-13 (pbk): 979-8-8688-0922-4   ISBN-13 (electronic): 979-8-8688-0923-1
https://doi.org/10.1007/979-8-8688-0923-1

Managing Director, Apress Media LLC: Welmoed Spahr
Acquisitions Editor: Celestin Suresh John
Desk Editor: Laura Berendson
Editorial Project Manager: Kripa Joseph

Cover designed by EstudioCalamar

Cover image designed by Freepik (www.freepik.com)

Distributed to the book trade worldwide by Springer Science+Business Media New York, 1 New York Plaza, Suite 4600, New York, NY 10004-1562, USA. Phone 1-800-SPRINGER, fax (201) 348-4505, e-mail orders-ny@springer-sbm.com, or visit www.springeronline.com. Apress Media, LLC is a California LLC and the sole member (owner) is Springer Science + Business Media Finance Inc (SSBM Finance Inc). SSBM Finance Inc is a **Delaware** corporation.

For information on translations, please e-mail booktranslations@springernature.com; for reprint, paperback, or audio rights, please e-mail bookpermissions@springernature.com.

Apress titles may be purchased in bulk for academic, corporate, or promotional use. eBook versions and licenses are also available for most titles. For more information, reference our Print and eBook Bulk Sales web page at http://www.apress.com/bulk-sales.

Any source code or other supplementary material referenced by the author in this book is available to readers on GitHub. For more detailed information, please visit https://www.apress.com/gp/services/source-code.

If disposing of this product, please recycle the paper

Table of Contents

About the Author ix

About the Technical Reviewer xi

Acknowledgments xiii

Introduction xv

Chapter 1: Economics of Ecommerce Business 1

Overview 1

Ecommerce Business Model 3

Myntra – Pure Marketplace Model 6

Marketplace Models and Profitability 7

Revenue Drivers 7

Cost Streams 10

Marketing and Promotions Cost 11

Technology and Platform Cost 12

Operations Cost 16

Economics of Ecommerce: Profit and Loss Statement 20

Summary 24

References 25

Chapter 2: Ecommerce Platform: Digital Ecosystem of Buying
and Selling 29

Overview 29

Browse by Category 32

Ecommerce Platform: Empowering Sellers Digitally 38

Top Funnel: Bring Visitors to Platform 44

Mid Funnel: Engaging Visitors with Products 51

Lower Funnel: A Path Conversion or Real Outcome 56

Summary...58

References 60

Chapter 3: Merchandising for Ecommerce Marketplace 63

Overview 63

Category Management...66

Case Study: Brand Prioritization on Ecommerce Platform 68

Site Merchandising in Ecommerce 73

Digital Marketing in Ecommerce...77

Summary...78

References 79

Chapter 4: Ecommerce Search – Powerhouse of Conversion 81

Overview 81

Search Queries and Machine Learning 84

Search Algorithms – Text Matching 91

Text Matching : BM25 Algorithm 92

Term Frequency (TF) 93

Inverse Document Frequency (IDF)..93

Search Result Ranking..95

Search Architecture in Action: From Query to Results 96

Semantic Search...98

Deep Learning for Search Embeddings 102

Conversational Search – Powered by Gen AI 103

Summary...105

References 106

Chapter 5: Curated Choices Using Art and Science of Recommendations 109

Overview 109

Recommendation Engines for Ecommerce: Engage Buyers with Curated Choices 111

Recommendation Engine – Business Impact..................................... 113

    The Science of Similarity: Crafting Personalized Choices 115

Recommendation Engine Architecture: Crafting Personalized Choices 118

    1. Data 119

    2. Similarity Measures... 120

    3. Algorithms 122

    4. Evaluation 130

Fashion Ecommerce: Recommendations for You 132

Similar Products – Personalized Product Recommendations............ 136

Neural Collaborative Filtering (NCF) 140

Content-Based Filtering Using Deep Learning 141

Style Up or Complete My Look Using Computer Vision 142

Conclusion 143

References 144

Chapter 6: Ranking: Science of Sorting in Ecommerce 149

Introduction.. 149

Ranking in Ecommerce 150

Ranking Function 156

    Ranking – A Brief History 158

    Google PageRank – Revolutionizing Search............................. 158

Search Ranking in Ecommerces 160

Ranking: A Deterministic Model.. 162

Ranking: A Machine Learning Model... 164

Ranking Algorithms: Learning to Rank (LTR) 167

Ranking for Recommendations on Ecommerce 169

Ranking for Similar Product Recommendations 171

Conclusion 173

References 174

Chapter 7: Personalization – AI-Crafted Customer Experience 177

Introduction 177

Location-Based Personalization 179

Home Page Personalization 180

Impact of Personalization in Ecommerce 184

Personalization in Marketing 187

Search Personalization 189

Design to Delivery Personalization: Stitch Fix – A Personalized Stylist 191

Personalized Similar Product Recommendations 192

Summary 197

References 198

**Chapter 8: Efficiency a Key Enabler for Delivery Experience
and Profitability 203**

Introduction 203

The Ecommerce Maze: Navigating the Order Fulfillment Journey 205

Efficiency Equation 208

Returns Orders 211

Customer-Initiated Returns (CIR) 212

Product Content: Creation and Validation 214

AI-Based Size Recommendations: Enabling Customers with
Right Decisions 215

Size Recommendations: Machine Learning (ML) Model 216

Size Recommendation – Skip-gram-Based Model 218

Recommending Clothes Sizes with Product Size Embeddings (PSE)..........221

Benefits of PSE 223

Reducing Customer Returns – Flagging Platform Abusers..............................223

Non-deliverable Orders and Return to Origin (RTO) Transactions 225

Cancellations 226

Summary..228

References 229

Conclusion **233**

Index..**241**

About the Author

 **Ramgopal Prajapat** is a seasoned Data Science and Artificial Intelligence professional with over 20 years of experience spanning financial services, ecommerce, media, and other industries. He has held leadership roles at Tata CLiQ, IBM, and Accenture, working with global clients across the USA, Australia, the UK, and India. At Tata CLiQ, part of the Tata Group, Ramgopal spearheaded machine learning initiatives focused on customer experience, recommendations, search, and transaction risk management. As an Associate Partner at IBM, he led transformative Data and AI programs across various sectors, including the Government of India, media, and asset management companies. Ramgopal's extensive experience with firms like Accenture, HSBC, and Infosys also includes co-founding a hyper-local ecommerce venture, where he developed an ecommerce platform from the ground up. His passion for teaching and his awareness of a significant gap in AI and marketplace understanding among professionals inspired him to author a book that provides comprehensive resources and actionable strategies for mastering AI-powered ecommerce.

About the Technical Reviewer

Prashanth Josyula is an enthusiastic expert in technical literature and a committed professional in programming. With a deep-seated passion for technology, he excels in crafting sophisticated code and exploring cutting-edge innovations.

Currently serving as a Principal Member of Technical Staff (PMTS) at Salesforce, he brings over 16 years of extensive experience and expertise in the IT industry. As a seasoned polyglot programmer, he commands a broad spectrum of languages and technologies, including Java, Python, Scala, Kotlin, JavaScript, TypeScript, Shell Scripting, SQL, and a variety of open source solutions. Since embarking on his software engineering journey in 2008, he has immersed himself in a diverse array of technological domains:

- Java/JavaEE, Spring: Architecting resilient and scalable backend systems

- UI Technologies: Designing intuitive user interfaces using ExtJS, JQuery, DOJO, Angular, and React

- Big Data: Leveraging Hadoop, Spark, Hive, Oozie, and Pig to extract actionable insights from vast datasets

- Microservices and Infrastructure: Engineering robust and scalable solutions with Kubernetes, Helm, Terraform, Spinnaker, and contributing to various open source projects.

- AI and Machine Learning: Navigating the frontiers of artificial intelligence and machine learning.

Each day in his career is an exhilarating journey of discovery and advancement, as he continually strives to redefine the boundaries of technological innovation.

Acknowledgments

I am deeply grateful to my daughter, Dr. Ankita Prajapat, who was the key inspiration behind this book. She encouraged me to share my knowledge through writing and suggested creating a YouTube channel – something I still plan to pursue. I also want to express my heartfelt thanks to my wife, Nirmala Prajapat, and my son, Girish Prajapat, for their unwavering support throughout this journey.

My time and experiences at Tata CLiQ were instrumental in shaping the ideas and insights shared in this book. I extend my deep appreciation to the leadership team at Tata CLiQ, as well as to my data science team there – Mahesh Gera, Anantharaman S, Prathamesh Fuldeore, Bhanu Pratap Singh Sikarwar, Niteesh Kumar, and Arpan Dutta Chowdhury – whose projects and discussions have played a role directly or indirectly toward the book.

Special thanks go to my colleagues at Tata CLiQ, Ruthraiah Thulasi, Biswajit Pal, and Avinash Nethrakere, for their invaluable feedback on both the content and writing style, which helped reshape this book.

I am also grateful to Suvadip Chakraborty and Gaurav Goyal for meticulously reviewing each chapter and providing pointed feedback that significantly improved the final product.

A special mention goes to Sanjay Ojha and Rajneesh Pathak for their consistent mentorship and support.

Finally, this book would not have come to life without the support of the Apress editorial team, who made the publishing process smooth and memorable. My sincere thanks to Celestin, Kripa, and Shobana.

Introduction

Ecommerce is an integral part of our daily lives as consumers. This is a thriving industry offering numerous opportunities for people seeking careers in the digital space. Whether you're a data scientist, product manager, category manager, or business leader, the ecommerce sector provides a dynamic environment where innovation is at the forefront and a key industry for the future of work.

In the digital-first world of ecommerce, the application of Artificial Intelligence (AI) is not just common – it's essential. From personalized product recommendations to dynamic pricing, AI and Machine Learning (ML) have become the backbone of modern ecommerce. While some applications of AI are well-established, new use cases are emerging at an unprecedented pace, driven by innovations in technology and the ever-evolving behaviors of the buyers.

However, for professionals transitioning from other industries into ecommerce, the learning curve can be steep. Industry-specific terminologies, key performance metrics, and unique business models can present challenges, even to seasoned professionals. This book is designed to bridge that gap, accelerating the learning process and equipping readers with the knowledge they need to succeed in ecommerce.

The recent rise of generative AI tools like ChatGPT has further fueled the desire to harness AI across various functions within ecommerce. While the enthusiasm is palpable, product managers, category managers, and business leaders often struggle with identifying the right AI and ML use cases and understanding how these technologies can deliver tangible returns on investment. This book addresses these challenges head-on, offering real-world examples and detailed outcomes of AI/ML applications, empowering decision-makers to confidently embark on AI-driven initiatives.

Identifying AI and ML applications in ecommerce requires more than just technical know-how; it demands a deep understanding of the business context and a clear vision of where to focus efforts for maximum impact. This book offers a non-technical, application-oriented perspective on how ecommerce businesses operate, how they create value for stakeholders – including customers, investors, and employees – and how AI and ML can be strategically implemented to enhance that value.

The first few chapters provide a foundational understanding of ecommerce business models, the key facets of ecommerce platforms, and the various stages of the customer journey. These are described in conjunction with the key performance indicators (KPIs) that drive success in ecommerce, and how AI and ML can influence these metrics to improve outcomes.

As the book progresses, it delves into the core AI and ML interventions that are transforming the ecommerce landscape. From advanced search algorithms and recommendation systems to personalization strategies that enhance the customer experience, the book covers the full spectrum of AI applications across the customer's journey. It also addresses post-order processes, where AI plays a crucial role in optimizing operations, managing risks, and driving profitability.

Whether you are new to ecommerce or looking to transform your business through AI and ML, this book is a comprehensive guide that will equip you with the tools and insights needed to succeed. It offers practical examples and actionable strategies that can be directly applied to your work, making it an invaluable resource for anyone looking to embed AI into their role, functions, or platform functionalities.

This book is more than just a guide – it's a companion for every ecommerce professional who aspires to stay competitive in the hyper-competitive digital world. The book is a prerequisite for any professional joining the ecommerce industry and can be used as a key resource during brainstorming sessions aimed at transforming ecommerce businesses into AI-embedded powerhouses.

Each chapter of this book is designed to be self-contained, allowing readers to explore topics in any order. However, the initial chapters lay a crucial foundation for understanding the ecommerce business model and should be read first. These chapters provide essential context and conceptual groundwork before diving into more specific topics such as search, recommendations, ranking, and personalization.

The next chapter offers a detailed examination of the ecommerce business model. It provides an in-depth look at the key drivers of revenue and cost within the marketplace framework and explains the intricacies of an ecommerce company's profit and loss statement. Additionally, the chapter highlights how AI and ML technologies can be applied to enhance profitability, offering valuable insights into how these advanced tools can optimize financial performance and support business growth.

CHAPTER 1

Economics of Ecommerce Business

Overview

While the buzz surrounding ChatGPT, a generative pre-trained transformer chatbot, has captured headlines, overall artificial intelligence (AI) and machine learning (ML) have quietly made significant strides across numerous industries. AI/ML models are empowering doctors with disease diagnostics, enabling ecommerce platforms to review products at scale and flag fake ones, and assisting banks with automated credit decisioning, showcasing the widespread applications of AI/ML. The automated credit decision engine is often powered by a machine learning model, typically developed using logistic regression due to its ability to explain the factors contributing to credit decisions – whether to accept or reject an application – but is not limited to this technique.

Ecommerce, a digital native industry, is rapidly evolving into an AI-powered factory. Algorithms are not only personalizing product recommendations based on browsing history and past purchases, but also influencing how products are displayed on the home pages and search

© Ramgopal Prajapat 2024
R. Prajapat, *AI-Powered Ecommerce*, https://doi.org/10.1007/979-8-8688-0923-1_1

results. Each step of the buyer journey in ecommerce is influenced by algorithms and data for improving experience and delivering business outcomes at the end.

Myntra, a leading fashion ecommerce platform in India, is utilizing AI to develop private label designs based on current trends and enhance buyer experiences. This includes personalized store fronts and an AI-Powered Styling Assistant.

Amazon India (amazon.in) is at the forefront of leveraging AI and ML from recommendations to address correction. Due to its scale and popularity, Amazon India faces challenges related to fake products and reviews, which are also getting addressed using AI and ML.

To create immersive shopping experiences and reduce the risk of returns, large retailers like IKEA offer buyers Augmented Reality-(AR) enabled applications. These apps help buyers in visualizing furniture in their homes before making a purchase, enhancing the overall shopping experience for the buyers and improving conversions (visitors placing orders). Augmented Reality (AR) and Artificial Intelligence (AI) form a powerful synergy, particularly in crafting immersive ecommerce experiences. For instance, AI-driven algorithms like collaborative filtering can accurately recommend personalized products, while AR technology enables buyers to visualize these products in their own spaces.

Ecommerce has irrevocably reshaped modern life, and its influence is particularly pronounced in India, fueled by a burgeoning young population and widespread mobile adoption. The Indian ecommerce market has experienced exponential growth [5], poised to reach a staggering US$ 163 billion by 2026, according to IBEF industry report [19]. This trajectory is projected to persist, culminating in a market value of US$ 300 billion by 2030.

The convergence of ecommerce and artificial intelligence, two of the most significant innovations of our time, has revolutionized how we shop and unlocked a wealth of value for all stakeholders – society, employees, and business owners. This value creation in ecommerce business is driven by various business models from open marketplace model to direct selling business model.

The next section dives into the two major ecommerce business models. We'll explore how each model delivers business value, then dissect the key revenue and cost drivers specific to the marketplace model, additionally providing a roadmap to profitability using artificial intelligence and machine learning algorithms. For instance, personalized product recommendations driven by deep learning can significantly enhance customer experience and revenue. Reinforcement learning-based dynamic pricing can optimize conversions and margins, while machine learning-powered classification models can effectively identify fraudulent orders, reducing costs. These are just a few examples of how AI can be a catalyst for profitability in the ecommerce realm.

Ecommerce Business Model

Before diving into the transformative power of AI/ML in ecommerce, let's first lay the groundwork by exploring the key factors that influence profitability in this dynamic landscape of ecommerce. In India, the business to consumer (B2C) space is dominated by two main business models: direct selling (Inventory Model) and marketplace model. Figure 1-1 depicts the common type of B2C ecommerce business models. A business model shows how multiple parties like buyers and sellers come together to enable a transaction [14, 15].

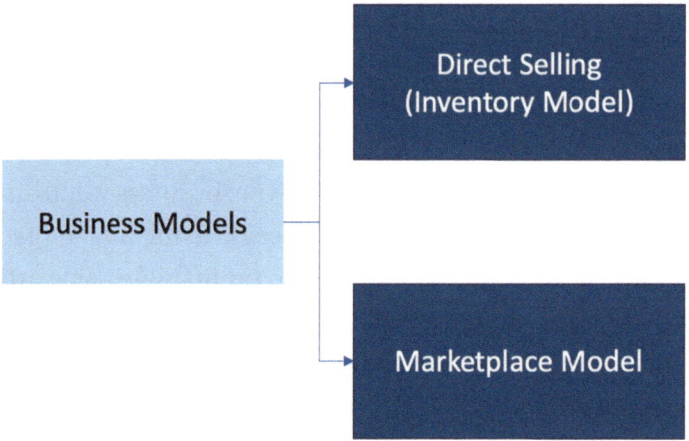

Figure 1-1. *Common E-commerce Business Models*

In the direct-selling business model, inventory, delivery partnerships, and customer service are owned by the companies who also own the ecommerce platform. They sell branded or private-label products through their websites and mobile applications. Some of the well-known examples of direct selling business models are Lenskart, an Indian ecommerce selling Eyewear products, and Nykaa, a leading Indian Beauty & Wellness product platform.

In the marketplace model, the ecommerce platform acts as a matchmaker, connecting buyers and sellers on a centralized platform. Unlike direct selling, they don't own inventory [2], but instead, earn revenue through commissions on successful transactions and by selling advertising space to brands and sellers. This model allows fashion platforms to offer a vast selection of products while keeping operational costs relatively low. Common examples of ecommerce platforms that have adopted the marketplace model are Flipkart, a multi-category marketplace owned by Walmart and Myntra is a Fashion ecommerce marketplace [3], owned by Flipkart.

Some of the ecommerce platforms have adopted a hybrid of these two – marketplace but hold inventory for some categories or own some private labels. This hybrid business model opens up a vast array of applications for machine learning and artificial intelligence. For instance, these platforms often employ multiple forecasting models to accurately predict demand, optimizing inventory management and warehouse resource allocation. Additionally, recommendation systems based on collaborative or content-based filtering, coupled with personalized experiences, are common across all ecommerce business models.

Understanding the core business model is a first step toward getting a deeper appreciation of key revenue and cost drivers. This involves exploring everything from buyer acquisition and conversions to logistics and operational efficiency. It's in this intricate interplay of factors that artificial intelligence and machine learning (AI/ML) contribute, offering powerful tools for optimizing every aspect of the ecommerce value chain.

While some of the large Indian ecommerce platforms have ventured into owning inventory and launching private labels, this chapter and the rest of the book will primarily focus on the powerful marketplace model. This allows us to fully explore the unique challenges and opportunities that AI/ML unlocks in this dynamic ecosystem of marketplace ecommerce.

AI/ML algorithms optimize everything from order allocation to personalized recommendations based on collaborative filtering or deep learning-based algorithms, for marketplace platforms are some of the common applications of machine learning algorithms. An order allocation, a critical component of ecommerce operations, involves determining the optimal fulfillment location for each order when products are available across multiple stores or warehouses. This process requires careful consideration of factors such as inventory levels, shipping costs, and buyer preferences to ensure timely and efficient order fulfillment. Machine learning, particularly classification and unsupervised learning techniques, plays a pivotal role in enabling informed order allocation decisions. But, it's important to acknowledge that additional use cases and considerations might emerge for other business models. For example, inventory management is crucial for direct selling business models and the quality demand forecasting models are critical for the direct sellers.

Myntra – Pure Marketplace Model

Myntra is a leading Fashion Ecommerce platform in India. Myntra Pure Play Marketplace (MPPM) is a fulfillment model [3]; in this marketplace model, a seller pays fixed fees, commission, logistics fees (forward and reverse) and charges for buying Myntra branded and mandated packing material. When a customer places an order on the Myntra app or website, the order is forwarded to the seller via order management system (OMS) integration. Once the seller packs the products, a pick request is placed to the delivery partner by Myntra and the order is picked up for the delivery. The order fulfillment is the responsibility of Myntra and typically fulfilled by its delivery partner – Ekart Logistics [4]. Figure 1-2 illustrates the working model of the pure play marketplace model at Myntra.

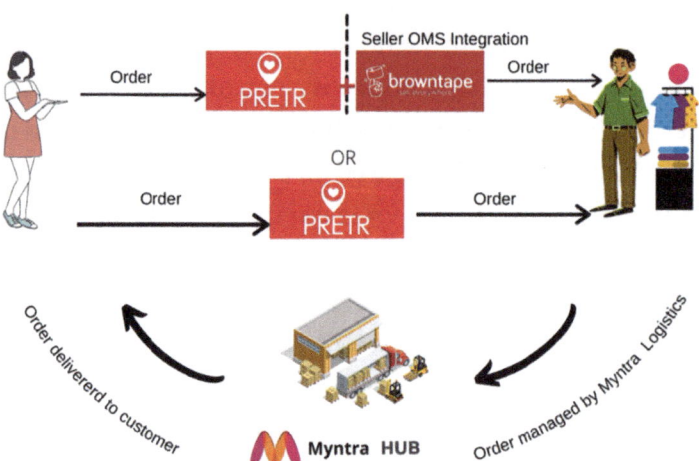

Figure 1-2. *Myntra Marketplace Model*

Source: https://browntape.com/what-is-myntra-ppmp/

The seller is also provided Seller Portal, where the seller is able to upload products (catalog), update discounts, get access to operational details such as order and fulfillment performance reports and financial reports like settlement and commission details. This is a sell-side platform in marketplace ecommerce business and discussed at length in the subsequent chapters.

Marketplace Models and Profitability

The success of marketplace platforms relies on a delicate balance between generating revenue and managing costs. For these platforms, the primary revenue stream comes from commission that is charged to brands and sellers. Considering the significant daily volume and monthly active visitors, sellers are also incentivized to leverage the platform's advertising functionality by purchasing ad space and that is an important revenue lever for these ecommerce platforms.

Revenue Drivers

Common revenue drivers for marketplace models include commission fees (like Amazon and Flipkart), advertising charges (like Amazon), and platform fees (like Myntra and Tata Cliq) [1]. Additional revenue streams can include listing fees and delivery fees (like Flipkart and Amazon).

The structure of these charges may evolve or change over time and across buyer segments. Figure 1-3 shows the common revenue drivers in the marketplace ecommerce business model.

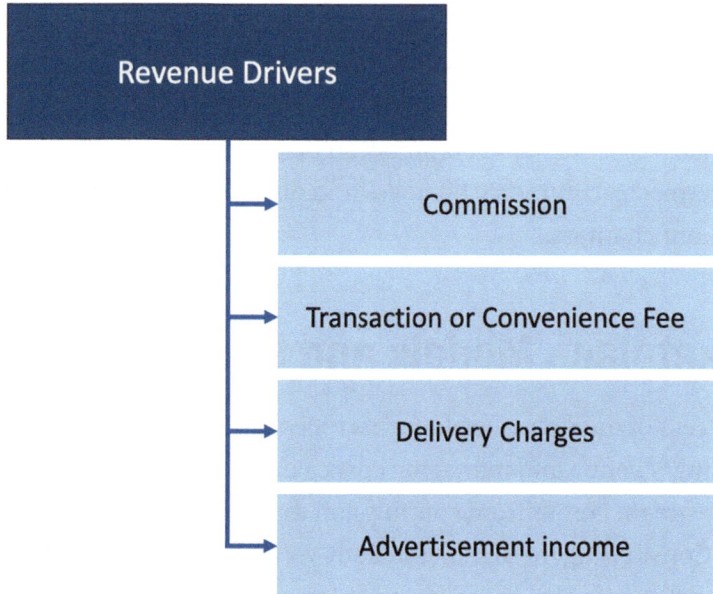

Figure 1-3. *E-commerce Revenue Drivers*

When a buyer completes a transaction on a marketplace platform, the platform itself earns revenue through a commission. This commission is a percentage of the total transaction value, often referred to as Gross Merchandising Value (GMV). The commission rate charged to sellers can vary depending on the industry, product category, and specific marketplace platform. For example, Nykaa reportedly charges sellers a 25% commission on the item price [12], while Myntra's commission rates range from 20% to 35% based on the product category [13]. Some platforms like meesho does not charge any commission to the sellers [10].

Some of the platforms charge delivery fees to the buyers for each product in an order and also levies convenience and/or platform fee. These are additional transaction-based revenue drivers, though these are not directly connected to the transaction amount.

"Myntra, India's second biggest ecommerce site, has implemented a Rs 10 charge on every order irrespective of the order value. Also, if you aren't a Myntra registered user and have an order worth less than Rs 1,000, then this is an additional charge to the existing Rs 99 fee"[11].

Beyond commission income and transactional fee/charges, the marketplaces generate revenue through advertising. These platforms boast millions of monthly active users (MAU), making them attractive destinations for brands and sellers seeking to promote their products. According to the report [16], Meesho has over 120 million average monthly active users (MAUs) and is growing at a rapid pace. This advertising income provides a significant additional revenue stream for ecommerce platforms.

These platforms maximize ad revenue by utilizing collaborative filtering or similar algorithms to deliver personalized ads that enhance click-through rates (CTR) and the ads revenue. Additionally, they employ reinforcement learning algorithms to determine optimal ad bids in real time. These platforms continuously balance exploration (learning from new data) and exploitation (maximizing current earnings) using multi-armed bandit algorithms for ongoing optimization.

For the selling of ad space on the platform, seller/brand can show banner ads, or the sellers can bid on keywords relevant to their products to ensure their offerings appear prominently in search results. Figure 1-4 below illustrates a real-life example of how an ecommerce platform displays sponsored products when a buyer performs a search.

Results

Price and other details may vary based on product size and colour.

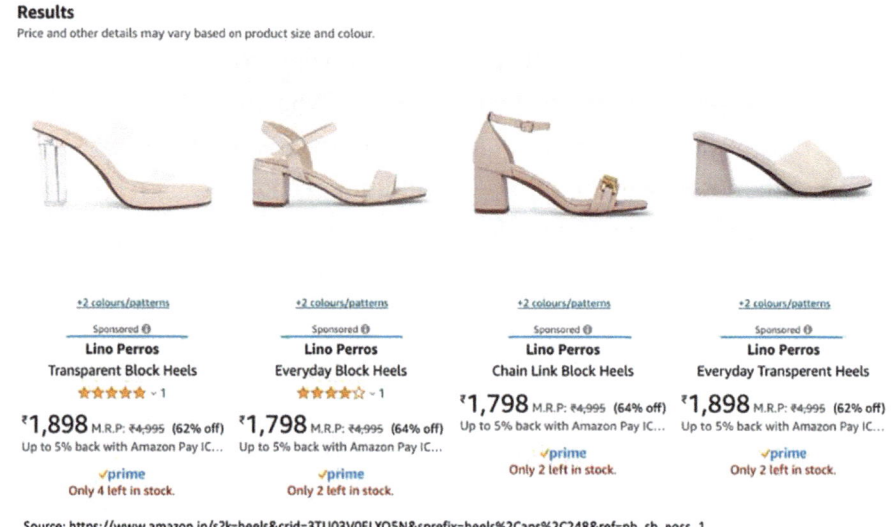

Source: https://www.amazon.in/s?k=heels&crid=3TU03V0ELXQ5N&sprefix=heels%2Caps%2C248&ref=nb_sb_noss_1

Figure 1-4. *E-commerce sponsored listing example*

Flipkart's Advertising revenue was at Rs 3324 crore in FY23. This is an increase from Rs 2083 crore in the previous period [8].

AI/ML algorithms, such as collaborative filtering and deep learning-based recommendation systems, play a crucial role in prioritizing and displaying personalized sponsored listings. By analyzing user behavior, purchase history, and browsing patterns, these algorithms can accurately predict user interests and preferences. This enables them to present ads that are highly relevant to each individual user, significantly increasing the likelihood of clicks and potential purchases [17].

Cost Streams

Ecommerce marketplaces face a multitude of costs across marketing, technology, and managing their operations.

The marketing expenses cover the cost of attracting new customers (e.g., cost of running social media ads, etc.) and reaching out to existing customers via SMS or WhatsApp campaigns.

Ecommerce platform is fundament to marketplace model, and maintaining a robust and scalable platform requires ongoing investment in tech infrastructure (servers, data storage, security, etc.), software licenses for various functionalities (ecommerce transactions, seller management tools, data analytics), and continuous development efforts (bug fixes, feature updates, security enhancements).

For running an ecommerce business, the operational costs involve managing logistics and delivery costs, providing services to the buyers by answering their calls and queries, managing seller operations from onboarding them to payment settlements, and preventing platform abuses or fraudulent activities.

Marketing and Promotions Cost

Ecommerce platforms fight tooth and nail for customer attention, with marketing and promotions often swallowing 10-15% of revenue [6]. Myntra's massive ad spend of Rs 1758 crore in FY23 tells the whole story [9].

One of the strategic decisions the marketing team involved in is to prioritize their spend across marketing channels – such as social search, search media, etc., for buyer acquisition and installs. Media Mix modeling approaches predict the impact of different marketing channels (TV, social media, search) on sales, optimizing marketing spend and improving expected ROI.

For existing buyers, AI driven segments based on buyers' demographics, purchase history, and online browsing behavior enable targeted and personalized customer relationship management (CRM) campaigns to deliver higher conversion rates and buyer engagement. Personalized CRM campaigns can have various objectives, such as sending complementary product email campaigns to recent buyers. For each buyer, based on their recent purchases, machine learning algorithms like association rules or matrix factorization can identify a set of products that are typically bought together. This helps in selecting and curating products for targeted email campaigns, increasing the likelihood of buyer response to these campaign campaigns.

Personalization for marketing campaigns can take many forms. As shown in the example above, this could mean recommending products to each buyer based on their recent purchases. Personalization can also be tailored to specific offers and rewards, preferred styles or brands, or even a buyer's preferred communication style or layout. AI and ML enable personalization at scale; for instance, Large Language Models (LLMs) can create personalized communications based on buyer personas. Meanwhile, supervised machine learning algorithms like Random Forests can identify buyers with a high probability of responding to a particular campaign construct.

Technology and Platform Cost

The technology platform is the backbone of a successful ecommerce marketplace. It underpins the entire ecosystem, influencing every aspect of the buyer and seller journey. This platform fosters differentiation, shapes buyer experience, and drives both operational efficiency and profitability.

Technology platform cost is significant, encompassing development, maintenance, and the infrastructure powering the online ecommerce platform. However, these costs often don't rise proportionally with increases in Gross Merchandise Value (GMV) or buyer numbers.

Artificial intelligence (AI) and machine learning (ML) are emerging as powerful tools to manage and minimize these technology expenses. Here are some leading examples of how these algorithms are reducing technology costs for ecommerce:

- **Predictive Maintenance:** AI can analyze log data and other data sources to identify potential failures in servers and infrastructure components. This analysis can augment diagnostics and suggest preventive measures, improving infrastructure resilience. Machine learning techniques (such as Random Forest or BERT) can predict the likelihood of bugs in a given codebase, helping to strengthen the testing efforts before releasing the code to production.

- **Cloud Optimization:** Cloud platforms offer flexible, on-demand resource provisioning. AI algorithms can predict peak usage times and enable dynamic resource allocation, leading to cost rationalization. Server usage patterns are analyzed to predict these peaks and enable this dynamic allocation. For any ecommerce platform in India, running annual sales campaigns is common. Machine learning techniques can accurately forecast traffic at a granular level and the granular forecast can be used to scale cloud resources throughout the sales period.

- **Development and Testing Efficiency:** Developer and tester productivity significantly impacts technology cost-effectiveness. Large language models-(LLMs) enabled Generative AI tools can streamline development workflows, boosting developer productivity. Additionally, AI-powered testing tools can automate the testing process, reducing timelines and resources compared to manual testing. This expedites development, catches issues early, and reduces overall development and testing costs. LLMs models have been successfully applied across test case creation to program correction [18]. AWS has released a service called Amazon CodeGuru Security that helps identify vulnerabilities in code. This service leverages machine learning and automated reasoning for detecting potential security issues [20]. Similarly, there are multiple platforms or AI startups that help in improving development and testing efficiencies. These platforms use machine learning to automatically generate, execute, and maintain test cases.

The case study below illustrates a case study of improving platform monitoring with a proactive alerting system that is enabled with AI and ML.

Case Study: Proactive System Issue Alerting in Ecommerce

Context:

Ecommerce is a fast-moving, digital-native industry where platforms frequently release new features and improvements. These platforms often operate within a complex ecosystem of interconnected software and services. Any issues arising from these releases can have a significant impact on customer experience and revenue. For example, a bug in the checkout process, payment gateways, or product pages can lead to cart abandonment, directly affecting revenue and order volumes.

Given the potential for such disruptions, continuous monitoring of various business performance metrics is essential and requires an automated system.

Problem Statement:

Develop an intelligent monitoring agent that evaluates business performance parameters in real time and flags potential issues proactively.

Solution Approach:

The proposed solution involves a module that leverages machine learning methods to forecast the expected values of business performance measures at a granular, hourly level. These forecasts take into account factors such as the time of day, day of the week, and sales versus non-sales periods.

The system's core is built on the following components:

1. **Forecasting Business Metrics:**

   - A regression model is used to estimate key metrics, such as the number of orders or Gross Merchandising Value (GMV), on an hourly basis. These estimates are then compared with actual values for each hour.

   - Other machine learning techniques, such as Random Forests or deep learning methods like Long Short-Term Memory (LSTM) networks, can also be employed to predict revenue trends more accurately.

2. **Deviation Analysis and Alert Generation:**

   - The system continuously compares forecasted values with actual results. If there is a significant deviation, an alert is generated. The threshold for deviation is carefully calibrated to balance the investigation effort with the risk of missing a potential issue.

3. **Multi-Stage Issue Management Process:**

   - **Issue Identification:** The first stage involves detecting potential issues through significant deviations in business performance metrics.

- **Investigation Optimization:** Once an issue is identified, the next step is to reduce the investigation time. This involves analyzing system logs to identify specific system components or interactions that may be linked to the issue.

- **Rapid Resolution:** The final stage focuses on resolving the confirmed issue quickly. This includes automatically assigning tickets based on system mappings and sharing relevant parts of the log to accelerate the investigation process.

Outcome:

The implementation of this proactive monitoring system results in reduced business losses by identifying and raising alarms for potential issues early. It also leads to faster investigation and resolution of problems, minimizing the impact on revenue and customer experience.

Operations Cost

Ecommerce marketplaces thrive on streamlined operations, but managing numerous functional processes can lead to high operational costs. Some of the prominent processes are logistics and fulfillment, seller operations, category operations, and transaction monitoring for fraud and abuses. These processes cover every aspect of the buyer and seller journeys for an ecommerce platform.

However, the rise of Artificial Intelligence (AI) and Machine Learning (ML) presents an exciting opportunity to transform these operational processes, optimize costs, and enhance experience.

Indian ecommerce players have been spending 7 to 15% of its sales value (Gross Merchandising Value – GMV) on logistics [4]. Platforms like AJIO rely on third-party logistics (3PL), while giants like Amazon and

Flipkart (with its Rs 5789 crore logistics revenue in FY23 [8]) operate their own fleets. But regardless of the delivery model, AI/ML can transform these costly operations into profit drivers.

Here are a few examples of how AI/ML can help Logistics and order delivery and fulfillment operations.

- **Smart Order Allocation**: For marketplaces, a seller may be operating from multiple stores and each with different levels of performance, the order allocation engine can assign orders that improve customer experience and reduce delivery cost. For example, finding a store with low distance and reduced processing time. A machine learning-based ranking method can be used to assign ranks to eligible stores for fulfilling an order. The features considered in the ranking function may include the distance between the buyer's pin code and the store's pin code, delivery cost, estimated delivery time, store cancellation rate, and other relevant factors.

- **Product Return Reduction**: Product returns are a complex challenge in the ecommerce industry. Returns can occur for several reasons: product quality issues linked to specific sellers and products, fit and sizing problems often inherent to the ecommerce model where buyers cannot try products beforehand, and instances of abusive buyer behaviors. Each of these challenges requires tailored strategies to effectively manage and reduce returns. For identifying returns linked to abusive buyers, an ML-based predictive model can identify return requests before they occur, allowing proactive interventions. For the buyers, with higher fake and wrong return chances are

nudged to submit the return product images when
they visit the order cancellation pages. AI-powered
image recognition can verify return claims and
automate refunds, reducing fraudulent returns and
administrative costs.

- **Minimizing Cancellations and Fraud**: AI powered
  systems can help in preventing stockout and
  misplacement, leading to fewer buyer-initiated
  cancellations. ML models can analyze buyer behaviors
  and flag suspicious orders to prevent fraudulent
  transactions. In a real industry experiment, it was
  observed that simply displaying a message stating, "You
  may be charged a return fee for an incorrect return,"
  significantly reduced buyer-initiated cancellations,
  even without actually charging any buyers. However,
  this message should be targeted only at potential
  fraudulent buyers, as displaying it to genuine buyers
  could lead to dissatisfaction.

Beyond Cost Savings, the AI/ML helps in Real-time Delivery Tracking
and providing buyers with accurate and timely updates, enhancing
transparency and trust and reducing chances of them calling customer
service teams.

In ecommerce, the sellers' operations teams assist new and existing
sellers. They provide onboarding training, set up their online store, and
help new sellers to list their products. They review and approve product
listings for accuracy, adherence to platform policies, and compliance to
legal regulations.

Machine learning algorithms can automate product listing reviews,
improving consistency, and compliance without the need for extensive
manual intervention. AI-powered processes can also generate data insights
to help sellers optimize their operations and improve performance on the

marketplace. Additionally, augmented reality (AR) technology can create a more interactive and efficient onboarding and training experience for sellers.

Fraud and platform abuse prevention team focuses and monitors activities for fraudulent transactions, scam attempts, and other platform abuses. They coordinate across functional areas to set up and implement security measures to combat evolving threats and platform abuses.

Machine learning models play a crucial role in combating fraud and platform abuses. By analyzing buyer behavior patterns, these models can predict potential issues like return order fraud or cancellations with malicious intent. Real-time fraud detection is also possible through AI and ML algorithms, flagging suspicious transactions and protecting the platform and its users from financial losses.

The case study below depicts a framework for reducing platform abuses without much challenges to the genuine customers using machine learning models in addition to improving operational efficiency.

Case Study: I*dentifying Abusive buyers on an ecommerce platform*

Context: Ecommerce platforms often face the issue of fraudulent orders placed with no intention of receiving the product. This leads to increased shipping costs, negative seller experiences, and inventory inefficiencies.

Problem Statement: Reduce the costs associated with orders that are placed with melodious intent, identify repeat abusers to show a nudge before they proceed to order confirmation.

Solution: Due to repeated abusive behavior by certain buyers, the platform could introduce a nudge to notify buyers with a high risk of abuse that order returns may incur a return fee. A key challenge is that buyers often circumvent this by creating multiple identities using different mobile numbers if the mobile number is a mandatory and unique identifier. This can make it difficult to take action based on just a few orders without negatively impacting genuine buyers or causing unintended consequences for the platform.

To address this, feature engineering and network linkage analysis are essential. For example, it was observed that many returns were initiated from similar geographic locations. Implementing deep learning-based classification could help identify abusive customers, particularly when dealing with textual features such as names and addresses. Often, abusers use identical or random text for these fields, which can be detected through advanced analysis techniques. Outcome: Platform efficiency has net positive impact without even charging the return fee to any customer.

By automating repetitive and manual tasks, AI and ML-based processes free up valuable human resources to focus on strategic initiatives. This not only reduces dependence on manual review processes but also improves overall efficiency and availability of support services.

There are a wide variety of functional processes and areas that AI/ML algorithms help business in making better decisions and deliver improved outcomes. Quantifying the impact of the projects plays a crucial role in driving adoption and ensuring resources are allocated effectively. Connecting AI/ML project outcomes to specific components of the P&L (profit and loss) statement provides clear visibility into the financial benefits of these initiatives. This helps in prioritizing projects with the highest potential return on investment (ROI).

Next section covers the key components of the profit and loss statement of a typical ecommerce company.

Economics of Ecommerce: Profit and Loss Statement

Profit and loss metrics empower businesses to optimize performance, improve profitability, and make data-driven decisions. The key revenue metrics are Gross Merchandise Value (GMV) , Net Merchandise Value (NMV), and Contribution Margin (CM).

Gross Merchandise Value (GMV) is the total value of all items sold on the platform, including returns and cancellations, and **Net Merchandise Value (NMV)** is the GMV minus returns and cancellations. NMV/GMV is the measure of efficiency and can help find levers to improve the efficiency using AI/ML methods. How can we reduce returns or cancellations?

Order cancellations can be attributed to various factors, including price fluctuations on the same or competing platforms and unexpected charges discovered post-purchase. To mitigate these issues, transparent and consistent pricing throughout the order journey is crucial.

Fraudulent orders from repeat offenders can also contribute to cancellations. As previously discussed, machine learning can effectively identify and address such behavior.

Additionally, seller-initiated cancellations often stem from inventory or pricing discrepancies.

Managing product returns requires a multifaceted approach, considering factors related to the product, seller, and buyer. Implementing strategies to address these issues can significantly reduce the overall impact of cancellations on the ecommerce platform.

Marketplace ecommerce companies typically offer discounts on top of the prices updated by the seller team. The discount is one lever for improving sales and conversions. Percentage discount on the NMV is tracked by the team to measure and control the cost.

Gross Margin (if selling items as inventory model) or Gross commission (relevant for the marketplace ecommerce model) helps in finding the revenue for the ecommerce platform. The revenue from gross margin if gross commission is the main income or source of revenue in a typical ecommerce business. Other income consists of revenue from monetization/advertisement income, shipping/delivery fees and platform fees. Total revenue for a marketplace includes commission income and other income.

Total Revenue = Commission/Margin income + Other Income

Gross Profit = Total Revenue – Discounts

In generating the revenue, the ecommerce platform incurs the cost. The direct costs that can be attributed to the transactions are – logistics costs, payment gateway charges, COD charges, Buyers/Sellers compensation costs, and loyalty cost.

Contribution Margin 1 (CM 1) is net of Gross Profit and Direct Costs.

CM1 = Gross Profit – Direct Cost

There are a number of AI/ML initiatives that could be considered to reduce these direct costs. In a marketplace model, allocating orders/ transactions to right sellers and delivery partners can help in reducing the shipping costs (logistics cost). Similarly, optimized and intelligent processes reduce leakages that are considered under the buyer and seller compensation costs.

In an ecommerce platform, which acts as an intermediary between buyers and sellers, the platform often absorbs some losses and risks if it fails to identify issues and recover costs from the responsible parties. To manage such scenarios, the platform might allocate a contingency fund, which could be a percentage of its business revenue.

For instance, if a buyer places an order and then requests a return, the seller might claim the product was damaged and seek compensation from the platform. If the platform cannot provide sufficient evidence to determine whether the damage was caused by the delivery partner or if the buyer returned a damaged product, it may need to accept the seller's claim and compensate them accordingly.

AI-powered image recognition models can compare product images throughout the return process to verify claims of damage or defects. Additionally, patterns of repeated abuse can be analyzed to flag potentially fraudulent claims from sellers.

Ecommerce platforms spend 10-15% of their revenue [6] in acquiring new customers or engaging the existing customers. Brand marketing (BM), direct marketing (DM), customer relationship management (CRM), associated with engaging existing customers, costs are the key components of marketing cost. Not all the costs are variable or can be attributed to each transaction.

Performance Marketing (Direct and CRM Marketing Costs) is a variable cost and is considered for arriving at Contribution Margin 2.

Contribution Margin 2 (CM 2) = Contribution Margin 1 (CM 1) – Performance Marketing.

In the Contribution Margin 3, Brand marketing cost is considered.

Contribution Margin 3 (CM 3) = Contribution Margin 2 (CM 3) – Brand Marketing.

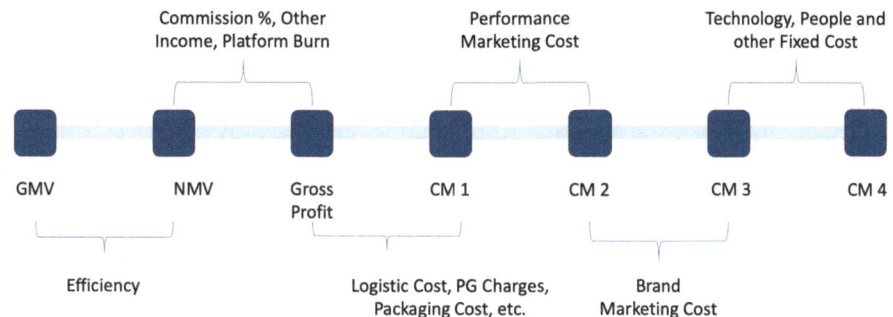

Figure 1-5. *E-commerce P&L Structure*

Again, AI/ML algorithms could play an important role in optimizing marketing decisions and improving returns on marketing investments.

Fixed costs including technology and employee costs are considered in finding Contribution Margin 4 (CM 4). As discussed in the previous sections, machine learning and artificial intelligence could help in optimizing and reducing various components of technology costs, leading to improved contribution margin 4.

Summary

Ecommerce businesses are built on technology platforms and have adopted various business models, especially for business-to-consumer (B2C) scenarios. The marketplace ecommerce model is one of the most successful business models, empowering a wide range of buyers and sellers. This model is linked to a set of revenue drivers and cost streams. AI/ML use cases can augment revenue levels and contain costs.

Transforming ecommerce platform sales into profitability is the cornerstone of the ecommerce business. Making informed decisions and strategic interventions not only boosts sales but also enhances overall profitability. By understanding the key components of the profit and loss statement of an ecommerce business, data science leaders can influence decision-making throughout the stages of ecommerce.

Throughout this chapter, we delve into the intricacies of revenue generation on ecommerce platforms and examine the essential costs embedded within the ecommerce business model. At each juncture, we investigate the various applications of artificial intelligence (AI) and machine learning (ML), which range from improving buyer experience to reducing platform abuses. In today's AI-driven landscape, while the potential of artificial intelligence is widely acknowledged, considerable skepticism exists regarding its return on investment. Measuring the return on investment (ROI) for each AI/ML project or use case can improve not only acceptability but also the profitability of the business.

By correlating the outcomes of each AI/ML use case with the profit and loss statement, we establish a clear framework for gauging and guiding investment decisions toward the AI/ML journey.

Building upon a comprehensive grasp of ecommerce business models, revenue streams, cost structures, and profitability factors, the upcoming chapter delves into the intricate mechanics of ecommerce platforms. This exploration spans both the buyer and seller experiences, with a particular focus on the customer journey – from initial engagement on social media to becoming a loyal customer on the ecommerce platform. Throughout this journey, we examine how machine learning (ML) and artificial intelligence (AI) are enhancing customer experiences, optimizing costs, and mitigating operational risks.

References

[1] Swagatika Simanchal Sabat, A Study of Marketplace Business Model in India, Mar-2021, `https://www.researchgate.net/publication/350107253_A_Study_of_Marketplace_Business_Model_in_India`

[2] Viktor Hendelmann, The Marketplace Business Model – A Complete Guide, `https://productmint.com/the-marketplace-business-model-a-complete-guide/`

[3] Browntape, BT Blog: What is Myntra PPMP?, `https://browntape.com/what-is-myntra-ppmp/`

[4] Myntra, Flipkart-Myntra Partnership Steps Up With Logistics Sharing, Jul-2025, `https://blog.myntra.com/merger-more-flipkart-myntra-to-share-logistics/`

[5] MARKET STUDY ON ECOMMERCE IN INDIA: Key Findings and Observations `https://www.cci.gov.in/images/marketstudie/en/market-study-on-e-commerce-in-india-key-findings-and-observations1653547672.pdf`

[6] Radhika Sridharan, Prashant Sarin, and Prashanth Aluru, India's Digital Fashion Disruptors, Sep-2023, `https://www.bain.com/insights/indias-digital-fashion-disruptors`

[7]   Siddharth Shankar, Flipkart Big Billion Days 2023: Inside Look at the Tech Driving India's Mega Sale, Oct-2023, https://www.timesnownews.com/technology-science/tech-behind-flipkart-big-billion-days-2023-article-104244002

[8]   BS Reporter, Flipkart Internet's operating revenue jumps 42% in FY23, loss declines 9%, Dec-2023, https://www.business-standard.com/companies/results/flipkart-internet-loss-reduces-9-revenue-increase-42-in-fy23-shows-data-123122800538_1.html

[9]   Soumyajit Saha, Fashion platform Myntra's FY23 revenue grows 25% to Rs 4,375 crore, loss widens https://economictimes.indiatimes.com/tech/startups/fashion-platform-myntra-sees-fy23-revenue-rising-to-rs-4375-crore-loss-widens/articleshow/106516005.cms

[10]   meesho, Pricing and Commission, https://supplier.meesho.com/pricing

[11]   **Rachna Manojkumar Dhanrajani**, Myntra introduces convenience fees for all, in a bid to become profitable, Jun-2023, https://www.businesstoday.in/tech-today/trending/story/myntra-introduces-convenience-fees-for-all-in-a-bid-to-become-profitable-384932-2023-06-09

[12]   Cointab, Nykaa commission on its marketplace, https://www.cointab.net/in/reconciliation-of-nykaa-marketplace-fee

[13]   Myntra commission charges, https://www.quora.com/
        What-are-the-listing-fees-for-Myntra-com

[14]   **Raphael Amit, Christoph Zott, Howard E. Aldrich,
        Charles Baden-Fuller,** Value drivers of ecommerce
        business models, Mar-2000, https://www.
        researchgate.net/publication/228556588_Value_
        drivers_of_e-commerce_business_models

[15]   **Peter Ractham, Ruth Banomyong, Apichat
        Sopadang,** Two-Sided E-Market Platform: A Case Study
        of Cross Border ECommerce between Thailand and
        China, Dec-2029, https://www.researchgate.net/
        profile/Peter-Ractham/publication/340933779_
        Two-Sided_E-Market_Platform_A_Case_Study_of_
        Cross_Border_E-Commerce_between_Thailand_
        and_China

[16]   TOI Tech Desk India's Online shopping market,
        How Amazon, Flipkart and Meesho are placed,
        Jan-2024, http://timesofindia.indiatimes.com/
        articleshow/107200100.cms

[17]   Yanwu Yang, Panyu Zhai, Click-Through
        Rate Prediction in Online Advertising: A
        Literature Review,2022, https://arxiv.org/
        pdf/2202.10462.pdf

[18]   Junjie Wang, Yuchao Huang, Chunyang Chen, Zhe Liu,
        Song Wang, Qing Wang, Software Testing with Large
        Language Models: Survey, Landscape, and Vision,
        Jan-2024, https://arxiv.org/html/2307.07221v2

[19]   IBEF, E-Commerce Industry Report, May-2024, `https://www.ibef.org/download/1720503483_E-Commerce_May_2024.pdf`

[20]   AWS, Amazon CodeGuru Security, `https://aws.amazon.com/codeguru/`

CHAPTER 2

Ecommerce Platform: Digital Ecosystem of Buying and Selling

Overview

Ecommerce platforms have revolutionized the way we shop, transforming brick-and-mortar stores into digital marketplaces. This is particularly evident in India, a country with a diverse population and a rapidly growing Internet user base. Ecommerce marketplaces have emerged as a powerful medium for connecting Indian buyers and sellers across the nation and beyond [3]. Overall, the eCommerce platform does more than just connect buyers and sellers; it facilitates comprehensive commerce through a cloud-based environment and includes a variety of additional features. Key features and functionalities of the eCommerce platform will be explored in detail later in this chapter.

Consider an example of Renuka, an IT professional from Mizoram, who works in Bangalore. She loves traditional clothing from Mizoram, especially for festivals and special occasions. She needs to find shops that sell similar products in Bangalore and this is not easy. Coming to the fast-forward world of ecommerce marketplace, Renuka can find a variety

© Ramgopal Prajapat 2024
R. Prajapat, *AI-Powered Ecommerce*, https://doi.org/10.1007/979-8-8688-0923-1_2

of traditional Mizoram dresses right up from her mobile and laptop. This is the transformation touching the majority of buyers in cities in India and across the world.

There are thousands of similar contexts and needs of buyers that are getting addressed via ecommerce platforms. Differentiation of ecommerce is more pronounced for specialized and/or hard-to-find products, e.g., organic tea in a small place or luxury branded products in smaller cities.

For an ecommerce marketplace to thrive, it must effectively support and enhance various aspects of the buyer's journey on the platform. This includes offering a wide range of relevant products, providing easy access to them, and ensuring a seamless process for exploring and completing orders.

For example, Renuka needs "Bandini Lehenga" and goes to an ecommerce platform and starts her shopping journey. She needs a way to reach the relevant products, this is facilitated by two ways – search what she needs or leverage category hierarchy (product taxonomy), or browse journey. When a buyer uses the search bar like in Figure 2-1, the platform shows the search results as listing form – see Figure 2-2.

Figure 2-1. *Ecommerce Search*

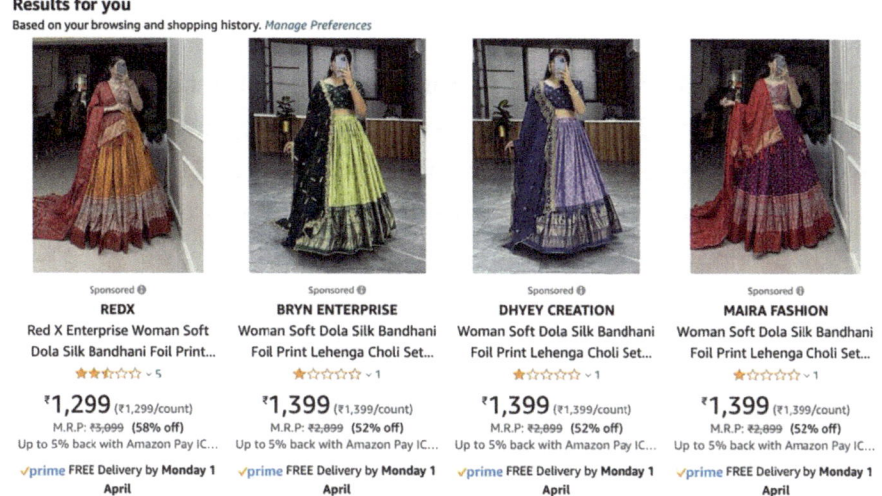

Figure 2-2. *Search result listings*

In addition to the search journey, a buyer can also reach to intended products using browse journey, as shown in Figure 2-3.

Figure 2-3. *Category Listing for Browse Journey*

Browse by Category

Search is a mechanism that maps a search query to products. This is typically enabled by ML algorithms. For category/brand browsing, the ranking algorithms enable showing popular, relevant, and personalized products on the top for buyers to have smooth browsing experience.

After reaching a set of relevant products, she is looking to explore additional products – similar designs and patterns before making a purchase decision. Buyer engagement or exploration of products are enriched by a wide array of recommendation features – frequently purchased together, "customers who bought this also bought" or recommendations for you or similar products.

In the offline Brick and Mortar shops, when Renuka goes to the retail shop (Brick and Mortar), a wide variety of products are displayed in the shop and a sales personnel assists her with additional products. The products are available for touch and feel for making sense of the quality and material relevant for clothing products. Also, trial rooms are available to try the actual products. Some of these needs are enabled by providing comprehensive description – covering product features, specifications, materials, and high-Quality Images and videos for the visual view of the products, enabling customers to feel how the product might look on her.

Considering touch and feel is taken away from the customers on the ecommerce platform, Artificial Intelligence and Machine Learning or Virtual Reality (VR) and Augmented Reality (AR) improve their online shopping experience. AR/VR try-on technology enables buyers to experience products in a simulated real-world environment, allowing them to see how spectacles will look. For example, when buyers are shopping for eyewear, it can show on their faces. Beauty platforms such as Sephora use facial recognition and similar AI technology to detect features like the eyes and lips, allowing customers to see how beauty products will look on their faces. Convolutional Neural Networks (CNNs), a type of deep learning method, play a crucial role in facial recognition by analyzing various nuances in images. For instance, when a buyer wants to try a particular shade of lipstick, the deep learning model identifies the lips and applies the selected lipstick color in real time. This allows the buyers to visualize how the lipstick will look on them.

IKEA, the world's largest furniture retailer, created an interactive digital experience for the buyers to see the 3D product experience of the furniture items in their own context and in real time using the immersive power of virtual reality [21]. In Virtual Reality (VR) applications, computer vision models play a crucial role in detecting and recognizing objects within the buyer's physical environment. When a new product is selected for placement, AI models extract features from the images to ensure the product fits accurately within the layout and optimizes its placement. This process enhances the realism and functionality of the virtual environment, helping buyers visualize how a selected product will look in their own spaces considering various other elements of the physical environment.

After discovering and selecting products on an ecommerce platform, it is crucial to engage buyers and guide them through the order completion process. Ensuring a smooth and appealing checkout journey is essential for converting interest into finalized purchases.

Offers and coupons are extensively used levers for attracting new buyers, incentivizing repeat purchases, and boosting sales [1]. They can also increase cart value by conditioning the offer on the number of items or order value. In 2023, 77% of the US population used coupons, and shoppers saved more than $3 billion by doing so [2]. Indian shoppers are similarly inclined to seek offers and promotions when making purchase decisions on ecommerce platforms.

The inefficiency or cost leakage due to lack of proper offer or coupon design could be significant. AI/ML and Analytics power the product bundling or designing offer conditions for maximizing return on offer cost. Buyer Segmentation empowers targeted promotions and personalized offers. The offers and coupons aligned with buyer preferences can increase engagement and conversion rates on the platform.

In a brick and mortar shop, payments are made only after receiving the product and cash is the mode of payment. A remarkable growth of ecommerce has seen many challenges on the journey. Firstly, a large segment of the Indian population still sees trust deficit in online transactions and access to the financial system remains a hurdle [8].

Cash on delivery (COD) is a significantly important payment option in ecommerce, with 5 out of 10 online transactions being done using this method. While COD provides convenience to buyers, it also presents challenges for ecommerce businesses, including a 35% higher product return rate and increased operational costs for cash handling [9]. Additionally, COD is linked to a higher percentage of products returned due to non-delivery, called return to origin (RTO) transactions.

To promote trust in online transactions and encourage digital payments, the Indian government has implemented initiatives such as "Digital India" and introduced BHIM (Bharat Interface for Money) to create a robust digital payment infrastructure. These efforts aim to reduce reliance on cash transactions and drive the adoption of digital payment methods in ecommerce. Cloud technologies are driving large-scale initiatives like Digital India by providing cost-effective solutions for delivering a wide range of services. These technologies enable efficient scalability, resource management, and accessibility, which are essential for the successful implementation and operation of such extensive programs.

For many online shoppers, the time between placing an order and receiving the product can be a source of anxiety. The uncertainty surrounding delivery timelines can erode trust and discourage purchases. The wait to receive and use the product also impacts the overall customer experience in ecommerce. Buyers often have concerns or uncertainties about the look and feel of the product. To address these concerns, the platforms offer end-to-end tracking and clear visibility about the order journey. Additionally, to alleviate concerns about product quality variation or delayed delivery/non-delivery, ecommerce platforms provide easy and fast refunds for faster purchase decisions.

Generous return and refund policies serve as a safety net for online shoppers, reducing the perceived risk of online purchases. Easy return policies and cash on delivery options have played a significant role in the adoption of ecommerce. However, they are also linked to a high return rate, which ranges from 25-40% in India, significantly higher than the global average [10]. AI/ML algorithms are helping these platforms combat abuse of these lenient return policies and maintain operational efficiency.

After an order is placed, several actions may occur, including order cancellations (by the customer or seller), product returns, or initiation of refunds (if the order was pre-paid). Buyers may also have various concerns or questions that they want to address with the ecommerce platform. Therefore, having buyer support available on the platform and over the telephone is an additional necessity for an ecommerce business.

A conversational AI can assist the customer service team in addressing customer requests. Certain requirements can be handled by conversational bots, which are available 24/7. Conversational chatbots employ Natural Language Processing (NLP) and machine learning classification models to categorize buyer inquiries based on their intent. Once the intent is determined, the chatbot searches for relevant information within the system and crafts a tailored response. This personalized approach ensures the buyer receives the most appropriate and helpful information. Moreover, AI and ML algorithms can enhance call volume forecasts and support staff requirements with higher accuracy, reduce dependency on manual and repetitive tasks, and enable intelligent process automation.

The call volume forecasting models can be developed using both statistical and machine learning approaches. Statistical models like ARIMA (AutoRegressive Integrated Moving Average) can capture linear trends and seasonal patterns in time series data. On the other hand, deep learning methods such as LSTM (Long Short-Term Memory) are well-suited for handling complex, non-linear patterns in time series data, making them a promising option for call volume prediction.

Ecommerce platforms leverage the power of user-generated content by displaying reviews and ratings from peer shoppers. This vast pool of peer-to-peer feedback plays a crucial role in influencing buyer decision-making. However, to ensure the credibility and value of these reviews, many platforms implement a review approval process before these go live.

Furthermore, some platforms go beyond basic reviews by utilizing Natural Language Processing (NLP) techniques. These methods allow for the creation of summarized reviews, highlighting key points, positive aspects, and relevant user tags. This concise format helps users quickly grasp the overall sentiment of user experiences for a particular product. Transformer models, including BERT, RoBERTa, and GPT-3, are essential for enhancing product reviews on ecommerce platforms. These advanced language models facilitate the generation of concise summaries and relevant tags, thereby significantly improving the overall user experience. Figure 2-4 shows how product reviews could be summarized and shown to the buyers.

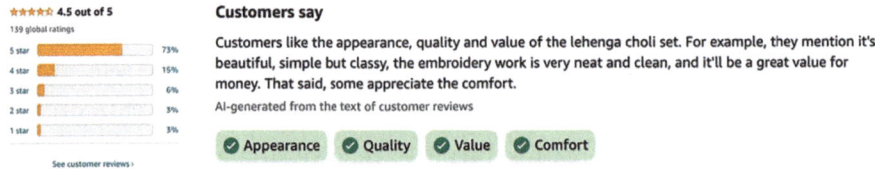

Figure 2-4. *Key highlights from reviews*

In summary, to facilitate the customer purchase journey on an ecommerce platform, key sections of a typical ecommerce platform include search and browse, offer and discounts, payment, and delivery information. This figure provides an overview of the fundamental components crucial for guiding buyers through their shopping experience on the platform.

Figure 2-5 provides an overview of the typical functionalities available to buyers on an ecommerce platform, illustrating the "Buy Side" features of the platform.

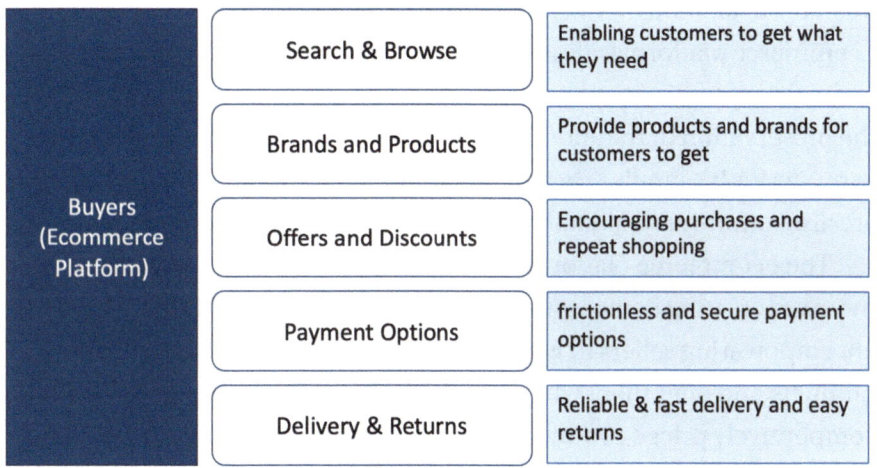

Figure 2-5. *Key functionalities of buyer-side ecommerce platform*

While search, navigation, product listings, reviews, recommendations, payments, and order management are all critical aspects of an ecommerce platform from the buyer's perspective, there are also essential functionalities working behind the scenes to ensure a seamless buying experience. These functionalities revolve around products and inventory management (such as listing products and ensuring availability), coupon configuration (determining which coupons are available and for whom), product pricing (managing product price updates), order management (assigning order processing responsibilities and communicating order journey information to relevant parties), and more. These elements are part of the seller side of the ecommerce platform, which will be the focus of the next section.

Ecommerce Platform: Empowering Sellers Digitally

Mehar is a talented handicraft maker in a small Indian village in Kerala. Ecommerce platforms allow him to sell his creations to a national audience, potentially attracting more customers and contributing to the preservation of traditional Indian craftsmanship and also get better earnings for his family. These are not isolated examples; millions of sellers are also empowered digitally by these ecommerce marketplace platforms.

The ecommerce platforms such as Flipkart, Amazon, and many more are creating democratized access for sellers across India. They are empowering sellers to get a vast volume of buyers to show their products and offer them what they do the best – better designs, creations, competitively priced products, or differentiated products.

These platforms provide all the technology functionalities (such as creating product listing, order processing, inventory management, etc.) for them to do business online without any worry of the technology changes or advancements. Even without any commitment or up-front costs. Most of these platforms take commission on the business values these sellers get from the platform.

These ecommerce platforms also enable sellers with a wide pull of audiences or buyers base across India, eliminating limitations of physical stores and reduced gestation period.

In addition, these platforms enable sellers to promote their products on-platform along with product level promotions (via sponsored listing) or they can promote their pages to social media platforms. These ecommerce platforms take care of delivery either via their own logistics fleet or via third party logistics partners, making it easy for the sellers to not worry about the actual delivery and fulfillment services.

Consider the example of Maya, a young woman, who is a jewelry designer. She has her own handicraft fashion jewelry shop in Jaipur, Rajasthan with the name "Maya's Mystic Jewellery." She wants to take her products to wider audiences and decided to be one of the leading marketplace platforms in India.

Maya registers on the ecommerce platform with all the supporting documents required for her to be a seller on the platform. Once she has registered her brand on the platform, she lists all the products on the platform.

Clear and concise product details such as title, description, relevant keywords, and attractive product visuals are required. A high-resolution photo from multiple angles, including close-ups to showcase details and variations, helps buyers in their purchase decisions. Some of the sellers take help of professional agencies to properly list products on these platforms. The platform also has validation and review functionalities to ensure that none of the fake or malicious sellers list their products to reduce risk for these platforms. The AI-enabled ecommerce platform assists Maya in generating product descriptions, creating product taxonomy, and filling in other details based on the uploaded product images. Additionally, computer vision-based methods can enhance buyer engagement by creating compelling images and visual content.

Considering the scale of these platforms, AI/ML solutions are a savior for them to manage product onboarding and review processes in check. Consider a case study of the open marketplace platform Meesho and how it manages product reviews on its platform.

Case Study: *Meesho Tackles Product Tagging with Deep Learning [12]*

Context: Meesho, India's third-largest marketplace (7% market share) behind Amazon and Flipkart [11], caters to millions of sellers, primarily from smaller Indian cities (Tier 3 and 4). Many of these sellers lack the technical expertise to properly categorize their products using the required product hierarchy (taxonomy). This makes it difficult for Meesho to display relevant products to users during searches and browsing.

Challenges: In the fashion category, inaccurate product details and a low attribute fill rate pose significant obstacles for a smooth buyer experience. Manually tagging products with a dedicated team presented scalability and consistency challenges.

Solution: Meesho's data science team has implemented a deep learning-based architecture to automatically classify product images into various categories. For products lacking attribute information, the system extracts relevant details, streamlining the product tagging process.

Outcome: An Artificial intelligence based scalable product categorization process improves the products taxonomy and data fill rate. Deep learning methods are highly effective at predicting product categories based on images and other provided details, thereby improving the fill rate. Users often make errors in selecting the correct category due to various pressures and may skip non-mandatory fields. Accurate product taxonomy and details are crucial for the search and browsing experience, so ensuring precise information enhances the overall buyer experience. This saves cost and also improves turnaround time.

Once products are on the platform, Maya should find a way to get business from these platforms as there are millions of sellers on the platform and some of the buyers may not be aware about the products from "Maya's Mystic Jewellery" brand.

Maya decides to leverage promotional and advertising functionalities to reach a wider audience to grow her online sales.

There are multiple functionalities typically available for the sellers on these platforms such as Sponsored Products. For a selected keyword, the seller sets a bid amount and pays to the platform based on the number of clicks on its product when shown to the buyers. This is a pay-per-click (PPC) advertising program. When customers search for those keywords, the products will be shown. The seller can promote their brand via Sponsored Brands program and it features the brand logo and other details in the top of search results. This creates higher visibility and brand awareness among the platform visitors.

Another option typically available is to show promoted products when customers are browsing related product categories, this drives traffic to the selected products and improves the purchase considerations.

Maya can come up with the target audience for her products and the list of relevant keywords. She can leverage brand promotion programs to create awareness about the brand among her target audience as this will help in creating brand equity and future sales. For immediate business results, she sets a daily budget target and bids for ten relevant keywords.

Sellers can get tens of orders for their products and if they can't deliver within the delivery timeline, it will create bad customer experience due to order cancellations. The customers will give bad ratings and there may be penalty charges for the seller from the platform.

The platform demands that sellers update the number of items available (inventory) on the platform so that it does not take more orders than the products available. Also, as soon as the buyer confirms the order, the order is available for the seller to do the next steps – pick up the product, package, and make it ready to hand over for delivery. This journey is typically available in the system called order management system (OMS). Some large-scale sellers with their own order management system work with the ecommerce platform to integrate these systems for faster and automated order detail handshake.

Case Study: Prioritized Seller selection for each product

Context: On an open marketplace platform, hundreds of sellers may be selling the same products. Showing the same products multiple times can lead to poor customer experience. Leading ecommerce platforms show a single product and pull pricing details from one of the prominent sellers.

*Approach: A fare- and performance-based seller selection is important for the marketplace as this has a direct impact on buyer experience and seller business. Using **seller performance** like average rating – higher rated seller gets higher weightage, time to process orders, **seller location** – a seller close to customer location will be prioritized due to lower days to deliver and reduced delivery cost, and **product details** like price (e.g., lower price is* *better for buyer) and inventory (higher inventory get higher importance) are considered for seller selection. These features along with machine learning algorithms help in finding the right seller for each of the products when multiple seller or seller stores are available. A common machine learning method for seller prioritization is ranking algorithms, which consider various factors to showcase products from top sellers.*

Outcome*: An automated and performance-based system to select the right seller for each of the products with multiple sellers or seller stores*

In addition to allocating order for processing to the seller, the system assigns order to courier partner. They may have multiple delivery partners (e.g., Blue Dart, Delhivery, XpressBees, Ecom Express, etc.) and AI/ML can enable the decision considering delivery timeline, logistic cost, and delivery performance features.

Promotions for "Maya's Mystic Jewellery" helps in getting the orders across India. She has a person who monitors the portal for the orders and processes these orders. Once the orders are packed and ready to ship, the information is updated on the portal. The orders are assigned to delivery partners and products are on the way to buyers.

Buyers receive the products ordered on the platform and are very happy with the quality of products and delivery timelines. They go on the platform and write product reviews and give excellent ratings (5*). These products start showing up in organic search and category pages in the commerce platform, thanks to the popularity ranking (Relevance Ranking) system. The relevance ranking (or Popularity Ranking) is powered by ML algorithms that consider multiple factors and create the ranking for arranging most relevant products at the top. Machine Learning-based Learning to Rank (LTR) techniques and Deep Learning-based ranking methods are employed to enhance the accuracy and relevance of products for the buyers.

Maya's online store – "Maya's Mystic Jewellery" is on the growth trajectory. She augments the product portfolio and also expands to other categories within the Jewellery segment. The main focus for her business is innovating new designs for her to grow the business. The technology functionalities and operational processes are taken care of by the ecommerce platform.

The money from the online sales is credited to the bank account automatically as per settlement timeline after deducting commission charges. For example, Nykaa reportedly charges 25% commission on the item price [13], while Myntra's commission rates range from 20% to 35% based on the product category [14].

These platforms also provide sellers with key insights and analytics to better manage their online businesses on the ecommerce platform. These insights range from sales metrics, such as order volume trends, to operational metrics, such as return rates.

In summary, to empower sellers on an ecommerce marketplace platform, a comprehensive range of functionalities is necessary, spanning from registration to financial settlement.

Figure 2-6 illustrates the functionalities available to sellers on an ecommerce platform, highlighting the "Sell Side" features of the platform.

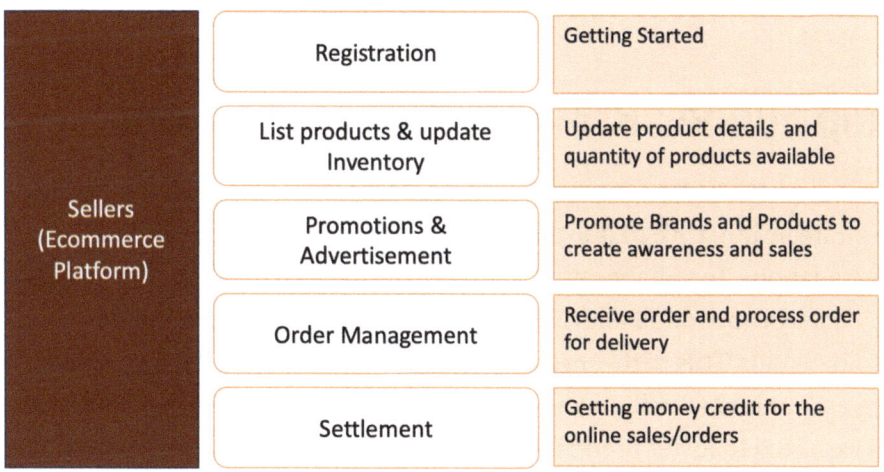

Figure 2-6. *Key functionalities of seller-side ecommerce platform*

In addition to these functionalities, sellers also need key performance statistics such as average order value, number of buyers or visitors, and conversion rate for their products. Analyzing these metrics helps sellers refine their strategies and enhance their business performance on the platform.

At the heart of every ecommerce platform lies a powerful promise: empower sellers and propel their businesses forward. To achieve this, the platform must become a well-oiled machine for attracting a steady or growing stream of visitors. Once they are there, it needs to captivate their attention with the right products, and entice them with strategic discounts and promotions. Finally, when a purchase is made, the platform ensures the chosen product reaches the customer swiftly and efficiently.

The upcoming section explores tactics for drawing visitors to the platform and steering them toward successful conversions. Buyer journeys on the ecommerce platform are segmented into three main funnels. The top funnel focuses on attracting visitors to the platform, the mid-funnel engages them through personalization and recommendations, and the lower funnel aims to convert visitors into orders.

Top Funnel: Bring Visitors to Platform

The marketing team at a leading ecommerce company has undertaken a strategic initiative to bolster brand recognition and drive mobile app installations. In collaboration with an external agency, they have produced a visually captivating television advertisement featuring its brand ambassador. This high-impact campaign will run for 30 days, reaching an estimated 2-3 million viewers daily. The TV advertisement campaign is expected to foster brand awareness and positive brand perception, increasing its mobile installs and visitors to their mobile application and website.

With large reach and mass appeal, brand building and emotional storytelling are some of the drivers behind ecommerce spending a large amount on the TV campaigns. But, TV advertisements face challenges such as lack of personalization, measurement of its effectiveness, higher cost with limited flexibility, and short attention span.

In addition to TV and print media advertisements, the digital campaigns are important tools available to the marketing team. A marketing team has a daily target to bring a certain volume of visitors and visits to the platform for achieving stiff growth targets. There are three segments of visitors on the ecommerce platform each day – first time visitors (new visitors), first time shoppers (repeat visitors, not shopped earlier) and repeat customers (shopped earlier). The targeted digital marketing campaigns are a measurable way to bring visitors across all three groups to the platform and help the marketing team deliver on the daily targets.

The digital performance marketing teams work with business/category and content teams to decide on the creatives, visuals, and offer constructs required for a digital campaign. Visual contents, format, and layout may be different across multiple digital marketing platforms. The visual content teams at leading ecommerce platforms have started using Generative Artificial Intelligence-(GenAI)based platforms (such as Midjourney or DALL-E) for generating visual contents for these campaigns.

Case Study: Enhancing Visual Content Creation with Generative AI

Problem Statement: Ecommerce platforms regularly run various sales and promotional events. As part of these activities, they frequently refresh the visuals on home pages and landing pages to align with promotional themes. Creating new imagery for these events is time-consuming and costly, posing a significant challenge for design and creative teams in ecommerce.

Solution Approach: The advent of Generative AI platforms, such as Midjourney and DALL-E, has revolutionized the process of creating visual content. These AI tools enable ecommerce design and creative teams to expedite the creation of refreshing visual content, significantly reducing the time and effort required. By leveraging these AI platforms, ecommerce businesses can quickly generate high-quality, themed images for their promotions.

Benefit: The use of generative AI significantly reduces the time required to create new visual content, allowing for quicker updates and more frequent refreshes. Additionally, it helps in reducing the cost of creating images due to the decreased manual effort, particularly in creating base images. Now, the main effort goes into review and final touch-ups.

Once appealing and attractive visual content is available, the digital marketing team decides to set up the campaign with a goal such as "conversion" for a social media campaign on a social media platform with limit on the budget to ensure the cost for the company does not go beyond the approved cost for the day.

The social media platform in the background runs experiments to assess the target users – who find this campaign attractive and relevant before scaling it. This is machine learning at play for the social media platform. How do social media companies use ML to show ads to the right audience? [15] Appreciation of these algorithms help the marketing teams in improving effectiveness of their campaigns and deliver better outcomes.

Depending on campaign objectives, promotional messages, and offer construct, the call to action may direct customers to either the home page, product listing pages (PLPs) or detailed product page (PDP). The data insights suggest the customers with entry pages as Product Detailed Pages (PDPs) have higher conversion rates as compared to when customers start with product listing pages (PLPs).

Conversion rate on an ecommerce platform is influenced by multiple factors such as buyer segment (may be influenced by campaign), level of discounts and promotions, assortment, personalization, and buyer journey on the platform. But it is important to bring buyers who have higher intent to purchase.

Social media campaigns bring measurable visitors to the ecommerce platform. The attribution of visitors to the campaign and source are done based on UTM (Urchin Tracking Module) parameters, allowing tracking of campaign source (e.g., Facebook ad), campaign medium (e.g., social media), and campaign name (e.g., "Spring Sale"). Facebook SDK (Software Development Kit) and pixels (often referring to Facebook Pixel) are commonly used for tracking and measuring conversions within ecommerce mobile apps [16].

An ecommerce company orchestrates a vast network of performance marketing campaigns daily or weekly across diverse channels like Facebook, Instagram, and YouTube. This strategic and multi-channel approach ensures they reach their target audience at every touchpoint.

The design of these campaigns goes beyond a one-size-fits-all strategy. Some campaigns focus on building brand awareness and loyalty. Imagine visually captivating content promoting Adidas products, perhaps featuring a special sale or a thematic campaign that aligns with the brand's identity. Additionally, they may run targeted campaigns for a select group of key brands across various categories, ensuring maximum visibility within the platform's extensive product selection.

A data-driven approach empowers the marketing team to systematically construct experiments and conduct A/B testing. This allows them to determine the most effective strategies for achieving a wide range of campaign goals, from app installs to conversions (purchases).

Social media and social search campaigns can sometimes have a higher cost per acquisition (CPA) compared to other channels. This is because the cost of conversion (measured by the return on ad spend, or ROAS) is calculated as the ratio of sales generated to the advertising budget spent on acquiring those sales. A well-crafted social media advertising strategy might achieve a ROAS of 4:1 (meaning for every $1 spent, they generate $4 in sales), whereas the average ROAS for Google Ads campaigns might be 2:1 [17].

Ecommerce companies extend their reach beyond social media and affiliate platforms by effectively utilizing their existing customer base. A key tool in this strategy is their Customer Relationship Management (CRM) program.

For app users who have opted-in for notifications, targeted push notifications deliver personalized offers and promotions. Additionally, the companies leverage segmented email and SMS campaigns to re-engage customers who have registered using their mobile number or email address. These targeted messages aim to pull buyers back to the platform, fostering a steady stream of daily visitors.

By employing a multi-pronged approach that goes beyond social media acquisition, ecommerce companies can effectively nurture existing customer relationships and drive repeat business.

While notifications offer a cost-effective way to reach buyers, bombarding them with irrelevant messages can backfire. A constant barrage can lead to notification fatigue, causing buyers to opt-out or even uninstall the app. Therefore, a well-planned, centralized strategy to manage buyer communication is crucial.

Imagine a typical ecommerce company with tens of category teams, each with daily targets. The temptation to send frequent notifications promoting their specific products becomes high. However, this disjointed approach can overwhelm buyers who receive a multitude of messages from different departments within the same app.

A centralized strategy ensures a coordinated approach where various category teams avoid bombarding the same buyers with redundant messages, leading to opt-out or even uninstall the app. Personalizing the content of push notifications can deliver a 4x lift on open rates, boosting the 1.5% average open rate seen with generic notifications to a much more respectable 5.9% open rate [18].

While SMS and WhatsApp messaging boast low per-contact costs, these advantages can be deceptive for ecommerce platforms with millions of registered users. Uncontrolled outreach can lead to significant expenses with minimal return on investment (ROI) due to low message relevance.

The key lies in data-driven personalization. A well-functioning CRM analytics and data science team empower the marketing team with AI/ML-powered insights. These insights predict which buyers are most likely to respond positively to specific product recommendations or promotional offers.

"Integrating AI into your CRM might create a more efficient sales process. Adding an AI-based e-commerce platform, like Customer Data Platforms (CDPs) or business intelligence (BI), will pave your way to personalisation, which will increase your average order value (AOV) and customer loyalty" [19].

While nurturing existing customers through CRM campaigns and performance marketing efforts focused on sales is crucial, relying solely on this approach can create a diminishing return. Over time, a natural customer churn (customers who stop using the platform) inevitably reduce the pool of returning and spending customers.

To ensure long-term growth, ecommerce platforms need a multi-pronged strategy. While retaining existing customers remains important, it's equally essential to continuously expand the app install base and explore new customer segments and acquisition channels.

Digital campaigns optimized for App Installs can attract new users and encourage them to download the ecommerce platform's app. Additionally, the team can look beyond existing marketing channels and demographics allowing the platform to tap into fresh customer pools. This might involve exploring new social media platforms, collaborating with different influencers, or targeting previously unexplored customer demographics with tailored messaging.

Ecommerce companies assess the performance of app install campaigns through key metrics like cost per install (CPI) and customer acquisition cost (CAC).

While cost-per-install (CPI) and customer acquisition cost (CAC) are crucial metrics for understanding the immediate cost of acquiring new users, they don't tell the whole story. An ecommerce platform might see a significantly higher CPI for their iOS app install campaigns compared to Android. However, upon analyzing post-install behavior, a key insight emerges. iOS users, despite costing slightly more to acquire, exhibit higher customer loyalty as measured by repeat purchase rates and average order value (AOV).

Customers who have had positive past experiences or hold a favorable view of the platform are likely to come and shop again, without direct results of any CRM or Digital campaigns. Word-of-mouth recommendations and brand campaigns on TV or digital channels could be other drivers of organic visitors on the ecommerce platform. Typical contributions of direct visits should be moving northward with the time for a successful ecommerce. Experience suggests that the direct visitors on an ecommerce platform could be between 20-40%.

Marketing teams play a critical role in attracting the right audience to the ecommerce platform within budget constraints. They achieve this through a two-pronged strategy. First through budget optimization and second by execution excellence. In budget optimization, the focus is to use data-driven tools like Market Mix Modelling, for strategically allocating budgets across channels like social search and social media and maximizing return on ad spend (ROAS) across all digital marketing campaigns and other marketing channels. On a daily basis, the marketing teams can strive for data and Artificial Intelligence-(AI)-based operational excellence by well-executed campaigns across social media and social search channels, ensuring they deliver targeted messages and a seamless user experience.

Complementing these efforts, the Customer Relationship Management (CRM) team and AI and machine learning (ML)-embedded CRM systems play a vital role in nurturing existing customers, fostering repeat purchases, and building a loyal customer base. By leveraging customer data and purchase history, the CRM can personalize recommendations, offer targeted promotions, and ultimately convert one-time buyers into loyal brand advocates.

Successful performance marketing, brand building campaigns, and CRM efforts all work together to cultivate customer awareness and interest. Once these initial stages attract visitors to the ecommerce platform, the conversion phase becomes critical. Here, the focus shifts to engaging these visitors and guiding them through the remaining stages of the marketing funnel.

Mid Funnel: Engaging Visitors with Products

A customer's initial touchpoint with an ecommerce platform can happen at homepage, product listing pages (PLPs), or Product Detailed Pages (PDPs). Visitors arriving through brand awareness efforts like TV commercials or direct visitors land to the homepage, whereas targeted marketing campaigns like paid search ads or social media promotions often direct users to relevant product categories or brands listing pages (PLPs) or specific products pages (PDPs).

The mix of visitors entering through these different pages depends on several factors, for example, marketing strategy, platform maturity, or brand awareness. The ecommerce platforms with a heavy focus on digital marketing might see a higher proportion of visitors entering through PLPs or PDPs due to targeted campaigns. Whereas established platforms like Amazon with strong brand recognition might see a larger share of visitors entering through the homepage due to direct traffic.

Understanding visitor source empowers ecommerce platforms to optimize the customer journey from each entry point. For visitors arriving through brand awareness efforts (e.g., homepage), prioritizing a visually appealing experience is key. This might involve stunning product visuals, engaging widgets, a clear search bar, and intuitive category navigation. To measure the effectiveness of the homepage, ecommerce platforms track the bounce rate – the percentage of visitors who leave the homepage without exploring further. Generally, a bounce rate between 20% and 45% is considered a good benchmark for ecommerce websites [20].

In the world of ecommerce, site merchandising teams play a critical role in capturing customer attention, especially on the crucial homepage. They act as the architects, meticulously configuring the layout and various elements that make up this digital storefront. The homepage is like an enticing façade for the platform and site merchandising teams carefully curate the key "widgets" that entice visitors to explore further. Some of the common widgets are Hero Banner, category navigation, or featured brand or offer widgets. Machine learning methods, such as Learning to Rank (LTR) and collaborative filtering, enhance category sequencing on home pages, thereby improving personalization.

The hero banner serves as the first impression, often showcasing seasonal promotions, new product launches, or captivating visuals and featured product or offer widgets highlighting specific products, special offers, or brand promotions, strategically placed to pique visitor interest.

Site merchandising teams are embracing the power of AI and technology to create a more engaging and personalized customer experience on the homepage. A/B testing remains a cornerstone of their strategy, allowing them to continuously optimize layouts and refine elements for maximum impact. Machine learning (ML) plays a key role in digital storefront design. Based on user clicks and purchases data, the algorithm can personalize the order of categories displayed on the homepage for each visitor. For example, if "women's kurta sets" have been trending recently, the algorithm might prioritize showcasing them first for visitors with relevant browsing and purchase history.

Imagine a homepage that adapts to each visitor's unique taste and preferences. AI can personalize various aspects of the homepage, for example, category sequences, featured brands, targeted offers, and curated product recommendations.

- **Category Prioritization:** Similar to dynamic category navigation, the most relevant categories for each visitor are displayed prominently.

- **Featured Brands:** Brands that a customer has interacted with or shown interest in can be highlighted on the homepage.

- **Targeted Offers:** The offer section can showcase deals and promotions from credit cards or stores the customer is more likely to use.

- **Curated Product Recommendations:** Leveraging past purchases and browsing behavior, AI can curate a selection of products specifically tailored to each visitor's interests.

These personalization tactics go beyond a one-size-fits-all approach, transforming the digital storefront into a unique curated experience for each customer. This level of personalization fosters deeper engagement and significantly enhances the customer journey.

After homepage, significantly high visitors land to category listing pages (CLPs) and brand listing pages (BLPs) due to performance marketing campaigns. PLPs both CLPs and BLPs receive visitors for a specific product category or Brand interest in mind. This targeted traffic can lead to higher conversion rates compared to the homepage, where visitor intent might be broader. The PLPs show a curated selection of products for a specific category or brand to shorten product discovery and expedite decision-making.

Showing the most relevant products on the top helps the buyers in making the purchase decisions faster. Ecommerce platforms leverage relevance ranking to prioritize the most relevant products on listing pages (e.g., PLPs, CLPs, BLPs). This is powered by machine learning (ML) algorithms. These algorithms use a variety of factors such as search results clicks, product views, cart addition rate (how often a product is added to the cart), conversion rate (how often a viewed product is purchased), and Gross Merchandise Value (GMV) and many more as input to ranking models.

For logged-in customers, the ranking goes a step further. ML algorithms can leverage customers' past purchase history, browsing behavior, and even abandoned cart products to personalize the order of products displayed. This level of personalization showcases items most likely to resonate with each individual customer, further accelerating their decision-making process.

Visitors on a typical ecommerce journey often encounter products that pique their interest on product listing pages (PLPs). This can lead them to click for a closer look at a specific product, landing them on the product detail page (PDP). Compelling visuals can entice visitors to click through to reach PDPs. Products on models (model images) and relevant background have typically higher click through rate (CTR) and user engagement.

Additionally, buyers can navigate from product listing pages to Product Detail Pages (PDPs). Paid search ads or SMS marketing campaigns may direct visitors directly to relevant PDPs.

Product Detailed Pages (PDPs) are the ultimate destination for purchase decisions in ecommerce. PDPs are dedicated to showcasing a single product and have a comprehensive overview of the product, showcasing the product from various angles with high-quality images, clear and concise descriptions that highlight the product's features, benefits, specifications, and materials along with displaying available sizes and colors.

PDPs must have clear Call-to-Actions (CTAs). The prominent CTAs include "Add to Cart" and "Buy Now" along with showing information around expected delivery date and customer reviews.

Figure 2-7 shows typically information shown on the detail product page (PDP).

Figure 2-7. *Product detailed page*

By providing a wealth of product information, engaging content, and clear Call to Actions (CTAs), the PDPs empower visitors to make informed decisions and ultimately convert into paying customers.

For measuring engagement on the platform, **product views per visit** (PV/V) – higher value indicate higher engagement of visitors in browsing and exploring product options, **Cart Addition Rate (CAR),** measuring percentage of visitors adding at least one product to the cart, and **time spent per session or visitor** – tracking average amount of time visitors spend across homepage, product page, category page, etc., and longer times indicate deeper engagement with the content.

Product Detail Pages (PDPs) are prime real estate and a well-designed PDP is for converting interest to a purchase. Multiple recommendation widgets are added to craft engaging PDPs. These widgets utilize data science and AI to personalize recommendations for each visitor. These recommendation widgets are Frequently Bought Together, Style It With, or Similar Products.

A well-curated product list with intuitive filters guides visitors to detailed product pages (PDPs). High-quality visuals, clear descriptions that highlight key benefits, and transparent delivery information on PDPs are crucial. These combinations effectively build trust and compel visitors to consider adding products to their carts. Once customers find the products of their choice, the goal is to convert these into a purchase.

Lower Funnel: A Path Conversion or Real Outcome

While adding products to the cart is a significant customer action, the real impact comes from order confirmation. These final steps in the ecommerce journey hold the most importance for a business's success.

In ecommerce, the average add-to-cart rate globally is 7.6% [4] and on average, ecommerce websites see a conversion rate of 1% to 4% [5]. The conversion rate can vary by industry and geography. For example, for the beauty category, the conversion rate is over 5%, but for luxury fashion it could be around 1% probably due to higher average selling prices (ASP) of the products. It is observed that the higher priced products take more visits before a purchase decision. **Add to cart rate** is calculated by dividing the number of visitors who have added at least one product to their cart by the total number of visitors to the product detail page. This metric assesses how effectively the product detail page generates interest and prompts users to select the product. **Conversion rate** is determined by dividing the number of visitors who complete the purchase after adding products to their cart by the total number of visitors who added items to their cart. This metric measures the effectiveness of the checkout and payment processes.

Ecommerce conversion rates thrive on a seamless checkout experience with multiple secure payment options (credit cards, debit cards, digital wallets), clear shipping information displayed up-front, and a streamlined checkout process with minimal steps. Pre-populated shipping addresses and saved payment information for returning customers further improve convenience, while offering a variety of trusted payment gateways caters to individual preferences and reduces cart abandonment.

Every buyer who abandons their cart during the ecommerce journey presents a valuable learning opportunity. These drop-off points can be analyzed to identify areas of improvement across UX/UI design (can we make navigation intuitive and easy and product information clear and easy to find?), product details(are product descriptions informative and engaging?), pricing and offers (are product prices competitive?), costs and charges (is fee and charges getting applied in cart pages?) and technical issues (is there any pattern of errors in the technical log?).

Marketing teams leverage CRM systems to reconnect with customers who abandoned their carts with unfinished purchases. Machine learning algorithms play a crucial role in these cart abandonment campaigns by personalizing the messaging, product recommendations, and overall content for each customer.

When a buyer is on the journey to purchase, inaccurate delivery addresses are a major cause of failed deliveries and lost revenue for ecommerce businesses. AI and machine learning algorithms can help mitigate this issue by offering address review services during checkout.

Payment is the final hurdle in the ecommerce journey. Offering a variety of secure payment options, including Cash on Delivery (COD), can streamline the checkout process and cater to diverse customer preferences.

In India, for example, the overall payment success rate is around 75% [7]. Interestingly, 33% of failed payments are simply not retried by customers [6]. By leveraging data and machine learning, ecommerce platforms can predict payment success rates for different options and priorities for each customer to improve first time payment success rate.

This prediction can be a powerful tool. Imagine informing buyers about payment methods with a higher success rate up-front. Additionally, prioritizing the most relevant and preferred options at checkout further enhances convenience. Previous experience suggests that the personalized payment sequence can improve usage of top payment options (an indication of customer experience) and also can help in improving adoption of pre-paid options. The payment method success rate for digital payments is currently less than 80%. By analyzing recent payment success rates, machine learning algorithms can offer better payment recommendations, potentially improving the buyer experience and increasing conversion rates.

Summary

The ecommerce platform empowers sellers with the technology they need to succeed. Sellers can list their products, run promotions, process orders, manage inventory, and receive secure online payments.

A marketplace ecommerce platform can host thousands of sellers. Ease of use, automated validations, and flexible options create core differentiations. Considering the scale, manual processes and linear cost growth pose significant challenges to survival for these platforms. This is where the role of Artificial Intelligence (AI) and Machine Learning (ML) becomes crucial. AI/ML-enabled processes make the platform scalable and cost-effective.

On the buyer's side, the platform offers a wide variety of products accessible from any mobile device or laptop, with seamless delivery to their doorstep. This book covers key functionalities and how these are enabled with Machine Learning (ML) and Artificial Intelligence (AI) to ensure a smooth and efficient experience for buyers. These functionalities include search, product listings, recommendations, personalization, and many more.

These platforms enable each stage of the funnel: top, mid, and lower. The key focus is to understand the applications of AI/ML across these funnels. The top funnel focuses on attracting visitors or buyers to the platform, which is primarily about marketing.

Machine Learning (ML) algorithms empower marketing teams to create targeted performance marketing campaigns that resonate with specific audience segments. Additionally, media mix models can optimize campaign spending and maximize return on investment (ROI) for the marketing budget.

On the platform, a personalized digital storefront tailors the shopping experience for each visitor. Relevance ranking and AI-powered recommendations ensure that buyers see products they are interested in, keeping them engaged and encouraging them to add items to their carts. This is the mid funnel, focused on engaging buyers on the platform.

A frictionless checkout process with secure payment options makes completing a purchase quick and convenient for customers. Personalizing the payment sequence based on buyer data and utilizing machine learning for address verification can significantly improve conversions and order placement. Converting visitors to buyers is the focus of the lower funnel in ecommerce.

With a detailed working overview of the ecommerce platform and buyer journeys on the ecommerce, the next focus is to delve deeper into each of the key functionalities. The next chapter explores how products, platforms, and promotions come together to engage customers on the platform, that is merchandising in ecommerce.

References

[1]   I. B. Suryaningsih, L. Farida, Ovilia Revanica, The Effect Of Coupon Sales Promotion, Online Customer Review And Perceived Enjoyment On Repurchase Intention In e-Commerce Shopee, Aug-2029, https://www.ijstr.org/final-print/aug2019/The-Effect-Of-Coupon-Sales-Promotion-Online-Customer-Review-And-Perceived-Enjoyment-On-Repurchase-Intention-In-E-commerce-Shopee.pdf

[2]   Somjit Barat, Lilly Ye, Effects of Coupons on Consumer Purchase Behavior: A Meta-Analysis, Jan-2025, https://www.researchgate.net/publication/298340203_Effects_of_Coupons_on_Consumer_Purchase_Behavior_A_Meta-Analysis

[3]   KPMG, Impact of e-commerce on SMEs in India, Oct-2015, https://assets.kpmg.com/content/dam/kpmg/pdf/2015/10/Snapdeal-Report_-Impact-of-e-Commerce-on-Indian-SMEs.pdf

[4]   Mckinsey, Why every business needs a full-funnel marketing strategy, Feb-2021, https://www.mckinsey.com/capabilities/growth-marketing-and-sales/our-insights/why-every-business-needs-a-full-funnel-marketing-strategy

[5]   Dynamic Yield, https://marketing.dynamicyield.com/benchmarks/add-to-cart-rate/

[6]   Adobe, Average ecommerce conversion rate benchmarks, Apr-2023, https://business.adobe.com/blog/basics/ecommerce-conversion-rate-benchmarks

[7]   Razorpay, Transaction Success Rate - what it is and
        why it matters, Dec-201, `https://razorpay.com/blog/`
        `transaction-success-rate-what-it-is-and-why-`
        `it-matters/`

[8]   ONDC and McKinsey, Democratising digital commerce
        in India, Apr-2023, `https://ondc-static-website-`
        `media.s3.ap-south-1.amazonaws.com/res/`
        `daea2fs3n/image/upload/ondc-website/files/`
        `Democratising%20digital%20commerce%20in%20`
        `India-Report-ONDC-McKinsey.pdf`

[9]   Indian Retailer, How eCommerce is affected by Cash
        on Delivery, May-2015, `https://www.indianretailer.`
        `com/article/multi-channel/eretail/How-`
        `eCommerce-is-affected-by-Cash-on-Delivery.a3422`

[10]   Rubal Sahni, Unlocking profit potential in retail returns,
        Apr-2023, `https://timesofindia.indiatimes.com/`
        `blogs/voices/207714/`

[11]   BUsiness Standard, `https://www.business-standard.`
        `com/companies/news/meesho-s-monthly-active-`
        `users-count-about-55-of-amazon-flipkart-`
        `report-123042101157_1.html`

[12]   Meesho, Automated Attribute Tagging Using Deep
        Learning Models , Aug-2022, `https://www.meesho.`
        `io/blog/tech-we-automated-attribute-tagging-`
        `using-deep-learning-models-data-science-`
        `machine-learning`

[13]   Cointab, Nykaa commission on its marketplace,
        `https://www.cointab.net/in/reconciliation-of-`
        `nykaa-marketplace-fee`

[14]   Myntra commission charges, https://www.quora.com/
What-are-the-listing-fees-for-Myntra-com

[15]   Facebook, Good Questions, Real Answers: How Does
Facebook Use Machine Learning to Deliver Ads?,
Jun-2020, https://www.facebook.com/business/
news/good-questions-real-answers-how-
does-facebook-use-machine-learning-to-
deliver-ads

[16]   Meta, About metrics for conversions campaigns.
https://en-gb.facebook.com/business/hel
p/997135723755052?id=1794272243992044

[17]   Improvado, Top 6 Strategies To Boost Your ROAS For
Social Media Ads, Aug-2024, https://improvado.io/
blog/6-ways-to-increase-return-on-ad-spend

[18]   Econsultancy, Personalisation can lift push notification
open rates by up to 800%, May-2026, https://
econsultancy.com/personalisation-can-lift-push-
notification-open-rates-by-up-to-800-study/

[19]   Elogic Commerce, How to Use Machine Learning and
AI in Ecommerce, May-2023, https://elogic.co/
blog/how-to-use-machine-learning-and-ai-in-
ecommerce-to-grow-your-business/

[20]   Adobe, Ecommerce bounce rate – what it is, why it
matters, and how to improve it, Nov-2022, https://
business.adobe.com/blog/basics/ecommerce-
bounce-rate

[21]   Demodern, IKEA Virtual Reality Showroom, https://
demodern.com/projects/ikea-vr-showroom

CHAPTER 3

Merchandising for Ecommerce Marketplace

Overview

An ecommerce marketplace brings together a wide range of sellers and buyers on a common technology platform. This platform facilitates product transactions and requires a comprehensive set of features for both buyers and sellers, as discussed in the previous chapter.

An ecommerce platform begins by identifying a set of target groups (TGs) to enable transactions. These target buyer groups represent the demographics and buyer segments that the platform aims to serve. Understanding the needs and requirements of these selected TGs is crucial for tailoring products and offerings for the buyers' preferences. Initially, when an ecommerce platform is new, target groups are defined based on market research and the founders' hypotheses. As the platform grows, machine learning methods become pivotal for identifying these groups by analyzing buyer purchase patterns and aligning them with buyer needs

© Ramgopal Prajapat 2024
R. Prajapat, *AI-Powered Ecommerce*, https://doi.org/10.1007/979-8-8688-0923-1_3

and personas. One commonly used unsupervised machine learning method for this purpose is K-Means clustering, which groups customers based on similarities in their behavior and attributes.

Ecommerce merchandising focuses on connecting buyers with the right products on the platform. Recommendation methods such as collaborative filtering and matrix factorization are commonly used to prioritize products for each buyer's target group based on their browsing and purchase patterns.

For instance, Myntra, a leading fashion ecommerce company, serves a specific target group, while another ecommerce platform, Amazon, is a multi-category platform that also sells fashion products. However, the brands and price ranges on these platforms are significantly different for the fashion category, and each of them targets different buyer segments.

Myntra targets fashion-conscious buyers, particularly within the medium-income group, who are looking for trendy and branded products [4]. The platform creates curated collections and focuses on providing exclusive products through partnerships with global brands. Its primary focus is on fashion categories across apparel, footwear, and accessories. Additionally, visually appealing design is a core focus for the platform [3].

Presenting relevant products to target groups (TGs) and anticipating their needs and preferences are key focus areas in merchandising. Data-driven approaches help map product needs to specific target groups. ASOS utilizes data to refine its target groups and align tailored value propositions to each group effectively [11]. Personalization and recommendations enable the presentation of relevant products to buyers. Monitoring emerging and evolving market trends, along with forecasting, provides insights for anticipating future product needs for each target group.

Merchandising in an ecommerce marketplace involves several key activities:

- **Curating Products:** Creating a curated selection of products from different brands and sellers and ensuring the availability of these products are the key priorities for the category managers in ecommerce. This is product category management. Product curation is a primary responsibility for category managers on an ecommerce platform. Machine learning-based recommendation and ranking methods assist them in selecting products that drive performance. Category managers continuously monitor key performance metrics, such as click-through rate, average order value, and conversion rate, to ensure the effectiveness of their product curation strategies.

- **Presenting Products:** Effectively presenting these products using attractive visuals on the platform and creating the right offer construct are the focus of site merchandising teams. Subtle variations in visuals can significantly impact buyer engagement on ecommerce platforms. The site merchandising team conducts various experiments to assess these effects, using statistical tests to measure their effectiveness.

- **Marketing Products:** Promoting products using enticing visuals, offers, and promotions across various marketing channels (marketing) to the right target groups and to bring them to the platform is among the functions of the marketing teams. Artificial Intelligence (AI) and machine learning (ML) methods are key enablers for personalizing marketing efforts, particularly for existing buyers. ML techniques such as XGBoost and Random Forest are commonly used to identify which customers are most likely to respond to CRM campaigns.

Ecommerce merchandising is a cross-functional responsibility that spans across category management, site merchandising to the marketing functions. Effective collaboration across these functions is essential to ensure a seamless shopping experience for buyers and to drive successful transactions on the platform.

Category Management

The availability of categories, brands, and products on an ecommerce platform creates differentiation for any ecommerce business. For a marketplace ecommerce platform, the presence of diverse brands and sellers defines its relevance and identity. In the fashion segment, the types of brands and price ranges indicate the target group that the platform aims to serve.

One of the core responsibilities of category managers is identifying brands and products that align with the platform's buyers. Creating a prospective list of brands and sellers involves understanding buyer needs, analyzing competitors to identify new opportunities and gaps in their product assortments, and distinguishing fashion trends and style innovations. Data and AI enable category managers to track the competitive landscape, capture buyer needs and preferences, and select appropriate trends that align with the buyers. Category managers typically monitor key competitor ecommerce platforms, including reviews and top-performing products and brands. Natural Language Processing (NLP) can analyze competitors' product descriptions, customer reviews, and social media mentions to provide actionable insights, helping them stay informed and make strategic decisions.

Stitch Fix is a US personal styling company founded by Karina Lake that uses machine learning algorithms and human stylists to deliver personalized fashion products to its buyers [5]. Beyond delivering "the right styles to the right people at the right time," the company has created a framework that helps in identifying future styles using algorithms. Some limitations of these algorithms are augmented by human insights and experience.

The process of generating ideas for new styles and deciding on investments involves a long gestation and iterative approach. The company tests new items via Style Shuffle even before producing and procuring them. This method improves the efficiency of producing successful styles and reducing waste [6].

Similarly, other leading ecommerce platforms like Zalando, H&M, and Alibaba are using AI for creating new designs, demand forecasting, and optimizing assortments. Google, Zalando, and the UK-based production company Stinkdigital have developed a predictive "design engine." This engine leverages AI technologies to enable creative decisions, showing color, texture, and style preferences of over 600 fashion experts [8].

After identifying buyer needs, emerging style trends relevant to the buyers, and a list of brands/sellers, the category managers engage the brands for onboarding to the ecommerce platform. The journey to onboard a brand involves multiple considerations before moving into operational and technology discussions. Identifying potential brands for an ecommerce platform involves both art and science. The scientific aspect is supported by machine learning methods and performance insights. For example, analyzing buyer preferences on the platform and key products from competing brands provides valuable data for brand selection. Additionally, Natural Language Processing (NLP) can enhance this understanding by evaluating product quality and fit based on buyer reviews from competitor websites.

Brand reputation and product quality are paramount for delivering a better experience for buyers on the platform. Typically, a platform has a set of guidelines and assessment criteria for onboarding brands that will deliver value to the platform's buyers. Exclusive partnerships or collaborations with brands enable platforms to offer unique products, creating a distinct edge over the competitors. The category teams evaluate whether the new brands contribute to a healthy mix of new, established, popular, and niche brands on the platform.

In an ecommerce marketplace, commission revenue is a critical component of the business model. The platform typically earns a percentage of each transaction value, which is negotiated and agreed upon by category management and the brand. Category managers play a pivotal role in driving this revenue by strategically managing product assortments and brand selections.

Case Study: Brand Prioritization on Ecommerce Platform

Context: In the rapidly evolving Indian ecommerce market, the identification and selection of brands for onboarding is critical to a platform's success. Using market research and competition analysis, the ecommerce marketplace identifies a potential list of brands. Some brands also proactively reach out to the platform for onboarding. A data-driven approach helps in systematically assessing these brands for alignment with the ecommerce marketplace's objectives.

Objectives: A data-driven framework for category managers to identify high performing brands, achieve a balanced mix of established and emerging brands and onboard brands that are aligned to buyer preferences and market demands. Tracking the performance of these brands and their products on competitor platforms, along with analyzing media reports, helps estimate brand performance and provides valuable insights.

Approach:

- Capturing data – sales, reviews and ratings, social media tractions metrics

- Creating key performance indicators (KPIs) such as sales potential, buyer satisfaction, and market relevance for each of these brands.

- Defining a scoring across these performance measures and aggregating the scores to arrive at a composite brand score.

- The composite score along with component scores guide category managers in evaluating and prioritizing brands for onboarding.

Outcome: The platform has implemented a structured approach to selecting and prioritizing brands for onboarding at regular intervals. This method has improved sales, product assortment, and the overall buyer experience by ensuring that the most relevant and high-potential brands are featured on the platform.

A long-term relationship with sellers and brands involves creating marketing plans, designing optimal pricing strategies, and developing promotional campaigns in collaboration with brands. These efforts aim to enhance brand visibility, increase buyer engagement, and achieve both the brand's sales plans (GMV) and the platform's revenue goals.

In an ecommerce marketplace, the brands or sellers are responsible for product inventory, availability, and order processing. Before launching a brand on the platform, a technical integration between the ecommerce platform and the brand's order management system is conducted to ensure seamless order processing. Inventory integration helps reduce the chances of order cancellations due to product unavailability. Flipkart, an Indian ecommerce platform. Leverages artificial intelligence in managing inventory [10]. Payment processing and analytics are other typical services and platform integration needs in the marketplace ecommerce model. Relevant agreements aim to reduce instances of product cancellations due to unavailability or poor quality, as well as delayed order delivery linked to slow packaging and handover to delivery partners.

In summary, category managers drive decisions around brand selection and product assortment on category and brand pages and responsibility of product availability on the platform. They also work to improve advertising revenue from brands and achieve sales targets (GMV and NMV) for the ecommerce platform.

Category managers track performance across funnels – top, mid, and lower – almost in real time to ensure they meet platform business objectives. Some common performance metrics are listed below. These metrics capture buyer journeys, identify opportunities to improve platform performance, and help in pinpointing any issues. For example, if certain visuals or widgets on the homepage are not effective or relevant to buyers, it will be reflected in the click-through rate of the widget. Similarly, if the conversion rate is lower in the previous hour compared to the expected conversion rate, the category managers will work with cross-functional teams to identify and rectify the issues.

For the top funnel, the volume of visitors, click-through rate (CTR) on the home and product listing pages, product views (PV), the ratio of product views to visits (PV/V), and bounce rate (% of visitors who leave without exploring beyond the landing pages) help in assessing the health of the funnel.

The mid-funnel focuses on buyer engagement, which is heavily influenced by product recommendations and personalization. Key metrics used to monitor buyer engagement include the cart add rate (% of visitors adding products to the cart), checkout rate (proceeding to the payment page), and the ratio of checkout (CO) to cart addition (CA).

Lower funnel performance is tracked using metrics such as conversion rate, order count, GMV, and NMV (considering cancellations), average order value, among others. These metrics help in making tactical merchandising decisions. For example, if the average order value is lower in recent hours, promotional strategies can be analyzed and adjusted. Additionally, product assortment and ranking may also play a role.

Figure 3-1 provides a summary of performance statistics for each stage of the funnels.

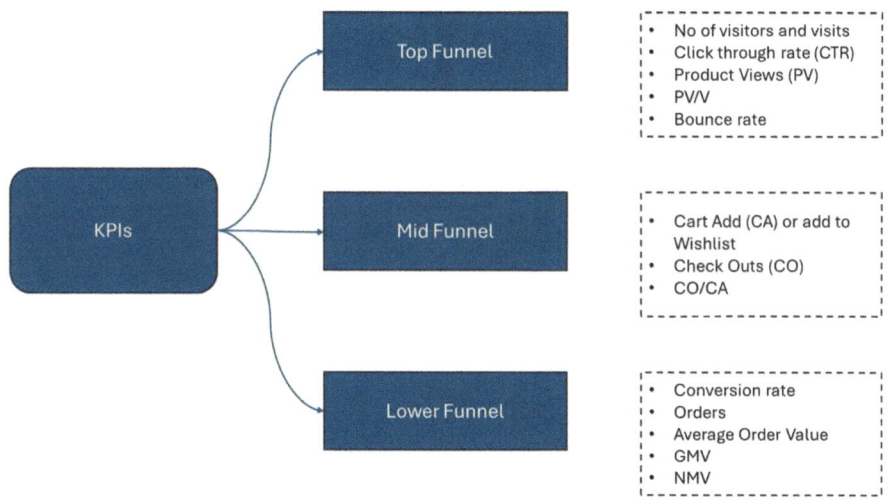

Figure 3-1. *Key performance indicators across marketing funnels*

Data-driven pricing and promotion with AI empower category managers to optimize sales and buyer engagement on the platform, balancing growth with profitability for the ecommerce platform.

On marketplace ecommerce platforms, some of the pricing decisions are left to the individual brands and sellers. The platform only controls the discounts on the top to drive the sales from time to time.

AI-enabled process controls can help ecommerce platforms to save costs and improve buyer experience. The case study below highlights impacts of operations control using AI for price and promotions.

Case Study: Price and Promotion Reviews using AI

Context: An Indian fashion ecommerce platform regularly undertakes price revisions and promotions to enhance sales and buyer engagement. With thousands of sellers, changes are made at an SKU level during sales periods. Discounts vary significantly across categories and brands, with some brands imposing limits on discounts while others offer substantial discounts seasonally. Managing these variations through rules across brands and categories poses a substantial operational challenge. Moreover, unintended pricing errors, such as a price changing from 999 to 99, can have significant financial impacts and occasionally lead to platform crashes during high traffic periods.

Problem Statement: Developing an ML model to review all price revisions and flag outliers before approving them

Approach: Using historical price revision data, we identified all price changes and flagged outliers through a combination of data analysis and business insights. Once labeled data was prepared, a machine learning model was developed to automatically detect abnormal price changes.

Results: The revised framework significantly reduced the need for manual review of price changes and accelerated the process of implementing price adjustments in production or going live.

Stockouts are a common challenge during ecommerce sales and promotions. Machine learning-based forecasting methods can help estimate demand by considering various factors such as discount levels, brand interest, and seller activity.

Regular operational engagement between brands and ecommerce platforms encompasses a range of responsibilities: from onboarding new products and enhancing product visuals, titles, and descriptions to monitoring demand trends, managing promotions, and addressing order management issues such as delayed packaging and increased returns.

High return rates represent a significant challenge for any ecommerce platform and can stem from various factors. These include product quality issues, inaccurate or outdated product descriptions, fit problems, or misuse by buyers exploiting platform policies. The category team actively engages with sellers to identify and mitigate these issues related to product quality and details on the platform.

Artificial Intelligence and Machine Learning is leveraged in review of the product details and approve products for onboarding for an ecommerce platform. Some of these models also flag fake and duplicate products.

With the right brands and products on board, the site merchandising team focuses on optimizing their placement and prioritization within the ecommerce platform.

Site Merchandising in Ecommerce

The site merchandising team plays a pivotal role in ecommerce, influencing buyer behavior and driving sales through strategic product presentation, placement, and promotions [1]. By analyzing buyer cohorts, purchase patterns, and emerging trends, the merchandising team designs layouts for home and category pages. They create widgets featuring specific product assortments and craft personalized buyer journeys.

Site merchandising enhances the overall buyer experience and boosts conversions by prominently featuring curated product assortments on home pages. This includes designing cross-category placements, establishing brand stores, orchestrating campaigns, scheduling promotions, and coordinating improvements to search functionalities. The team collaborates closely with category managers and marketing teams to align merchandising strategies with business objectives. They assign product assortments and promotions in alignment with category-specific trends and goals.

The featured widgets are designed to showcase curated product assortments, special offers, and promotional campaigns on home pages and specific category pages. The team continuously monitors the performance of these widgets and iterates on them to improve business outcomes. Key metrics tracked include click-through rates, conversion rates, and overall sales or Gross Merchandise Value (GMV).

Cross-category placements strategically position products from complementary categories to encourage buyers to explore a wider range of offerings. These combinations are designed using analytics and data science methods to optimize buyer engagement and sales.

In fashion ecommerce platforms, merchandising and category managers collaborate with brands to create brand stores and curate product assortments. These brand pages are promoted across various social media platforms to enhance visibility and drive traffic. Merchandisers work closely with marketing teams to schedule promotions for these curated assortments and brand stores, maximizing their impact.

The case study below explains details around how site merchandising enabled with AI/ML helped an Indian fashion ecommerce platform to improve business outcome.

Case Study: Personalized Curations for Regional Festivals

Context: India is a country rich in cultural diversity, with numerous festivals celebrated across its various states and regions. These festivals often have specific cultural significance and attract people from specific demographic segments, even those who may be spread across different geographical locations. For example, "Durga Puja" is a major festival celebrated with great enthusiasm by Bengalis, who are present in significant numbers across major Indian cities. The ecommerce platform aimed to leverage the cultural relevance and needs of these specific target groups by showcasing relevant products to the right audiences.

Objective: The goal was to create a curated collection of products relevant to Bengalis, a few weeks prior to the Durga Puja festival, on a fashion ecommerce platform.

Approach: The platform identified Bengali users based on their geographical location and behavior when they logged in. Utilizing computer vision and AI techniques, the team curated a selection of products that would be particularly appealing to Bengalis during the festival. These curated products were then showcased with captivating visuals and promoted through targeted marketing campaigns across social media platforms, specifically designed to attract Bengalis and related buyer segments.

At the core of the problem is identifying products similar to those typically purchased during Puja, particularly for the targeted buyer segments – Bengalis and northeastern buyers across India. Based on insights from category managers and domain expertise, relevant categories were identified. For these categories, data on products purchased by these segments 1-4 weeks before Puja last year was analyzed. A multi-model approach was employed to create product embeddings for each category. These embeddings were then used to match current inventory items. However, this approach has limitations, such as the inability to capture emerging trends. To address this, experts manually curated additional novel products.

Outcome: *The campaign led to a significant increase in new buyer acquisition on the platform, as well as improved conversion rates and sales for the curated products during the promotional period. To measure conversion rates, an A/B testing approach was employed. Bengali buyer segments were split into two groups: one group was shown the curated Product Listing Pages (PLPs), while the other group viewed the standard PLPs.*

India is a country rich in festivals, with celebrations occurring at regular intervals throughout the year. During these times, brands often introduce new product lines and curate items that are particularly relevant to the festive occasions. For example, "Diwali" – the festival of lights – is celebrated widely across India, especially in the northern regions. This festival is a prime time for shopping for clothes and gifts.

Fashion ecommerce platforms like Myntra and Amazon create curated collections specifically for festival shoppers and promote them extensively. The site merchandising teams also reinvent the platform's presentation or digital stores to align with the festival themes, aiming to attract and entice buyers.

In addition to creating festival-aligned brand stores, the site merchandising team collaborates closely with brands and category teams to orchestrate impactful end-of-season sales promotions. For example, an Indian ecommerce platform employs site merchandising techniques to prominently feature curated product pages, promoting end-of-season products across social media platforms. AI-driven behavior insights and data analytics enable them to create multiple pages for each brand, ranking products with the objective of improving conversions and sales. These pages are targeted to appropriate buyer segments through structured look-alike targeting campaigns. This strategic approach not only drives increased sales but also heightens buyer satisfaction by offering timely discounts on popular products.

By harnessing data-driven insights and fostering strategic partnerships with brands and category teams, ecommerce site merchandising teams optimize product visibility and sales performance during pivotal promotional events such as end-of-season sales. This iterative process empowers them to continually refine their merchandising strategies and achieve exceptional business outcomes.

The site merchandising team ensures accurate, comprehensive, and high-quality product information across the platform. This team acts as a bridge between the brand and the buyers, crafting a digital storefront that is both engaging them and enticing them to make purchase decisions.

Curated products with accurate and relevant product descriptions and high-resolution images and videos help buyers in making purchase decisions. The team works closely with brands/sellers, category managers, and product teams to gather the information and present products that are both visually appealing and easy to understand for the buyers.

Buyer behavior and engagement data helps in understanding interest and preferences of the buyers on the platform and these insights help site merchandising to implement strategies to enhance navigation, improve content relevance, and highlight promotions effectively. The team tracks a variety of performance metrics. These metrics include click-through rates, conversion rates, bounce rates, average session duration, and more. By analyzing these data points, the team can identify areas for improvement and implement targeted optimizations. For instance, if a particular category page has a high bounce rate, the team might refine the product listings or adjust the page layout to make it more engaging. This data-driven approach ensures that the site not only meets but exceeds buyers' expectations.

Merchandising focuses on presenting products attractively, while marketing involves promoting these products to target groups. This combination is essential for optimizing buyer engagement and conversions in the ecommerce marketplace platform.

Digital Marketing in Ecommerce

Ecommerce platforms extensively leverage digital channels and social media to attract buyers. Visual merchandising, promotions, and targeted messaging are key tools used to capture the attention of users on social media and motivate them to visit the ecommerce platforms. These strategies are increasingly data-driven, utilizing analytics to understand buyer behavior and preferences.

Visual marketing is essential for capturing the attention of buyers on social media platforms. The use of computer vision and artificial intelligence allows for a deeper understanding of how visual presentations influence buyer behavior. By enhancing visual content and developing innovative visual strategies, businesses can create a more engaging and cohesive shopping experience for buyers [12].

A data driven targeted approach not only enhances the effectiveness of marketing campaigns but also improves the overall shopping experience for buyers, leading to higher engagement and conversion rates. In the pursuit of data-driven marketing decisions, the ecommerce industry commonly employs lookalike modeling, programmatic bidding, and the monitoring and measurement of customer acquisition costs and return on marketing spend. These strategies help optimize marketing efforts and improve overall performance.

Summary

The category teams collaborate with brands and sellers to enhance product assortment and ensure product availability for targeted buyers. The site merchandising teams focus on creating visually appealing presentations, providing detailed product information, and ensuring a smooth purchase path, all of which contribute to a seamless shopping experience on the ecommerce platform. Marketing and promotional efforts are then employed to attract buyers to the platform, engaging and enticing them to make purchase decisions.

Personalized search and product recommendations offer a tailored shopping experience for each buyer, making it easier for them to find products that suit their preferences. Additionally, the use of ranking and sorting algorithms to prioritize various widgets based on shopping patterns further enhances user engagement on the platform.

In the next chapter, we will explore buyer engagement and the role of ecommerce search. This feature is a crucial functionality for ecommerce platforms, as it facilitates the product discovery journey by reducing the time and effort needed for buyers to find products that meet their needs and preferences.

References

[1]   Big Commerce, Effective Ecommerce Merchandising Tactics to Scale Online Stores, `https://www.bigcommerce.com/articles/ecommerce/merchandising/`

[2]   social media influencer, Unveiling the E-Commerce Giants: Amazon vs. Myntra, Dec-2023, `https://medium.com/@technologyrevolution/unveiling-the-e-commerce-giants-amazon-vs-myntra-d6d92c5b5c5a`

[3]   SAUMYA TEWARI, Amazon vs Myntra: Who will win over more Gen Z shoppers in India?, May-2023, `https://www.storyboard18.com/quantum-brief/amazon-vs-myntra-who-will-win-over-more-gen-z-shoppers-in-india-8247.htm`

[4]   Hitesh Bhasin, Marketing Mix of Myntra and 4Ps (Updated 2023), Jan-2024, `https://www.marketing91.com/marketing-mix-myntra/`

[5]   Stitch Fix, `https://en.wikipedia.org/wiki/Stitch_Fix`

[6]   VOGUE BUSINESS IN PARTNERSHIP WITH STITCH FIX, How Stitch Fix is using AI to predict trends, Mar-2024, `https://www.voguebusiness.com/story/events/how-stitch-fix-is-using-ai-to-predict-trends`

[7]   Stitch Fix, How We're Revolutionizing Personal Styling with Generative AI, Jun-2023, `https://newsroom.stitchfix.com/blog/how-were-revolutionizing-personal-styling-with-generative-ai/`

[8]   Achim Rietze, Project Muze: Fashion inspired by you, designed by code, Sep-2016, https://blog.google/around-the-globe/google-europe/project-muze-fashion-inspired-by-you/

[9]   ADGULLY BUREAU, Alibaba Empowers Fashion Retail with AI, Aug-2028, https://www.adgully.com/alibaba-empowers-fashion-retail-with-ai-79573.html

[10]   AIX, Case Study: How Flipkart is Leading the AI-Driven E-commerce Wave, Oct-2023, https://aiexpert.network/case-study-how-flipkart-is-leading-the-ai-driven-e-commerce-wave/

[11]   Aditya Shastri, Exclusive Marketing Strategies of ASOS – A Case Study, Jun-2021, https://iide.co/case-studies/marketing-strategies-of-asos/

[12]   Linmeng Liang, Visual marketing in e-Commerce applications analysis, 2020, http://proceedings-online.com/proceedings_series/SH-SOCIALS/ICSHS2020/emss01182.pdf.pdf

CHAPTER 4

Ecommerce Search – Powerhouse of Conversion

Overview

Jose, an avid runner, who frequently participates in Bangalore's events, needed a new pair of shoes. Like countless others, Jose turned to the web to research his next purchase. Statistics show a significant portion of buyers use online searches to decide on the products before visiting ecommerce platforms [1]. Knowing exactly what he wanted, Jose typed "Adidas Men Trainers Running Shoes" into the search bar of a popular Indian ecommerce platform, ready to begin his shopping journey. Figure 4-1 illustrates an example of a search bar on an ecommerce website. A similar search bar is also available on the mobile application for ecommerce platforms.

Figure 4-1. *E-commerce Search*

R. Prajapat, *AI-Powered Ecommerce*, https://doi.org/10.1007/979-8-8688-0923-1_4

Ecommerce search plays a critical role in these platforms, allowing users to navigate through billions of products and find what they need. It's a crucial contributor to sales, generating the largest percentage of transactions compared to other channels [2]. Typically, on fashion ecommerce platforms, 20–30% of buyers utilize search at least once per session, yet remarkably, they contribute a significant portion (50-60%) of overall sales (GMV). These search-savvy buyers often have a clear purchasing intention, leading to a conversion rate 2–3 times higher than the platform's average conversion rate. In addition to its impact on conversion rates and overall value, an effective search function creates a seamless and engaging buyer experience on the ecommerce platform. Therefore, search is a crucial component of the platform's strategy and user engagement.

As Jose hits enter after typing "Adidas Men Trainers Running Shoes" in the search bar, the ecommerce platform's search algorithms spring into action. The search algorithms match the search query terms with product title and descriptions and show the list of products that contains the search terms. The method is called exact text match.

Search functionality in ecommerce has come a long way. It has evolved from basic text matching to advanced semantic search, now powered by generative artificial intelligence.

An ecommerce platform might offer millions of products. Comparing every single product title and description to Jose's query term ("Adidas Men Trainers Running Shoes") would be incredibly inefficient. This approach would lead to very slow search response times, significantly impacting the customer experience. Since Jose's query clearly indicates he's looking for shoes, so no need to check other categories like apparel for matching the products.

Ecommerce platforms rely on product category hierarchies, also known as taxonomies, to organize millions of products into clear and distinct groups. These taxonomies are crucial for browsing and navigation. However, search queries themselves often don't neatly fit into these predefined categories. One approach would be to categorize searches based on keywords. But managing rule-based manual categorizations becomes complex as customers use a wide variety of search terms and they typically use informal language for the search queries [3].

To overcome these limitations, ecommerce platforms leverage machine learning-powered query classification models. These models act as intelligent classifiers, analyzing search query terms like "Adidas Men Trainers Running Shoes" and the model predicts the most relevant product categories and brand information. This allows the search algorithm to efficiently match Jose's query with the most relevant products in the catalog, significantly improving search response times and the overall user experience.

Figure 4-2 illustrates the high-level process of linking search terms to catalog products.

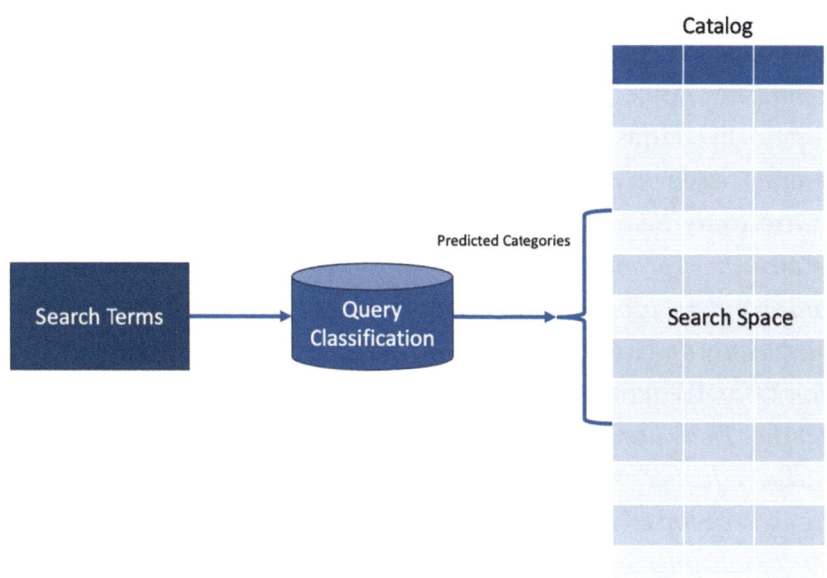

Figure 4-2. *Linking search terms to product catalog*

Search Queries and Machine Learning

It is important to note that search queries are often brief, with nearly 45% containing fewer than three words. This observation is based on the author's review of search data from an Indian ecommerce platform during a specific period. This suggests a significant proportion of buyers might be browsing or exploring the platform rather than having a clearly defined search intent. As a result, query classification models may encounter limitations in terms of accuracy for these shorter queries. Even for brief search queries, BERT and other transformer-based deep learning models can effectively classify search terms into relevant product categories. These models excel because they utilize not just the words themselves but also the semantic meaning behind the queries.

Additionally, ecommerce platforms leverage in-session clicks and previous searches to refine and improve search results, including query classifications. For example, if a user frequently clicks on electronics when searching for "Apple," the system may infer that the user is referring to the brand. In contrast, a search for "custard apple" would be categorized differently based on contextual clues.

Case Study: Search Term Classification Model

Context: *An ecommerce platform has three levels of taxonomy to organize apparel products. For example, Level 1 is Apparel/Footwear/ Home/Beauty and each of the Level 1 (e.g., Apparel) will have Level 2 (e.g., Indian Wear, Western Wear, etc.). Each of the L2 will have multiple Level 3 categories. There were 3000+ Level 4 combinations.*

All search queries that are not direct brand or category searches, were required to be linked to the taxonomy or category hierarchy to improve product matching.

Objective: *Develop a transformer-based text classification model that links each of the search queries to the relevant category hierarchy. This is a multi-label classification task.*

Approach: For the normalized or processed search queries, the category labels were assigned based on the click data. If most of the products clicked are related to a single category level, assign them as the correct level. High level logic used is explained below.

For example, if a search query is linked to over 30% of the clicked products related to the single Level 4 value, only that Level 4 and corresponding Level 3, Level 2 and Level 1 are considered for the label. If clicked products are across multiple Level 4, then only Level 3 will be considered.

This process of finding 30% of clicked contributions is started with Level 4, but repeated at Level 3 and upward. If over 30% of the products are clicked for a particular Level 2, then that level 2 and corresponding L1 will be assigned for the label. The process is repeated for Level 1 in the similar way.

Tshirt For Women – > Level 1: Apparel: Level 2: Women Level 3: western wear Level 4: T Shirt

Woman Dresses – > Level 1: Apparel: Level 2: Women

Once the label dataset is created, a transformer based pre-trained model is used for developing the text classification. But the dataset was skewed for some of the most shopped or searched categories. To make the dataset balanced, based on the product catalog, a list of synthetic search queries were created for the categories with lower count of organic search queries.

Model Validation: The trained query classification model was validated on the hold out sample and precision, recall, and F1 score values were used for performance evaluation.

While Jose knew exactly what he wanted and entered a specific search query, it's important to remember that buyer personas based on ecommerce search behavior vary greatly.

In contrast to Jose's targeted search, consider another user, Rashmi, a homemaker planning for an upcoming function at her apartment. She needs a saree but wants to explore various options before making a purchase decision. Unlike Jose, Rashmi represents a significant portion of ecommerce customers who are decided on a product category but seek a wider selection within that category. To fulfill her need, Rashmi utilizes a more exploratory search query: "Orange Sarees Price less than 2000." This search reflects her desire to browse a variety of orange sarees within a specific price range.

While a query classification model can effectively route Rashmi's search query ("Orange Sarees Price less than 2000") to the "Saree" category within the broader "Apparel" section, challenges remain. Product titles and descriptions might not always explicitly mention the price or color a user has in mind. Accurately capturing price intent within text matches can be particularly difficult.

Beyond classifying search queries, query understanding is crucial for ecommerce platforms. This involves extracting key entities like brands, prices, colors, and other attributes. This process, however, can be challenging due to the varied nature of search queries.

Natural Language Processing (NLP) comes to the rescue with a powerful technique called Named Entity Recognition (NER). NER acts like an intelligent decoder, identifying and extracting relevant entities from search terms. In fashion ecommerce, NER models could be trained to recognize a wider range of entities critical for product searches, including product type (like "saree"), brand name, color (like "orange"), material, gender, occasion, and style.

Figure 4-3 illustrates the different entities identified within an ecommerce search query.

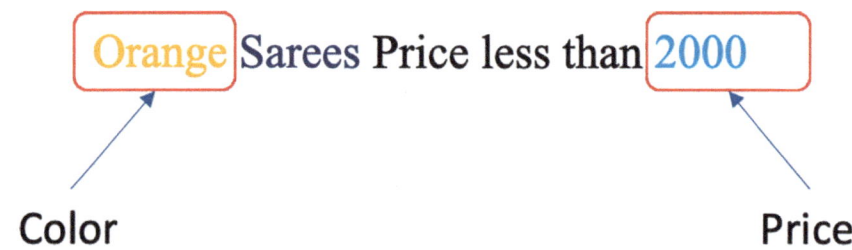

Figure 4-3. *Extracting entities from search query*

For ecommerce search, deep learning-based pre-trained models are commonly leveraged to build highly effective NER models [4]. Named Entity Recognition (NER) models, such as BERT, are trained on custom datasets specific to ecommerce to enhance their accuracy and effectiveness. Preparing these custom-labeled datasets can be time-consuming, but it significantly improves the search process. For instance, a word like "Apple" might be identified as a brand in some contexts and as a category in others. Additionally, certain word variations may refer to brand names rather than spelling corrections

In this scenario, color as "Orange" and price as less than 2000 is extracted. Search algorithms use these attributes for applying filter conditions before sharing the matched results to the user – Rashmi.

Figure 4-4 illustrates the sequence of Named Entity Recognition (NER) and classification models used to map search terms to catalog products.

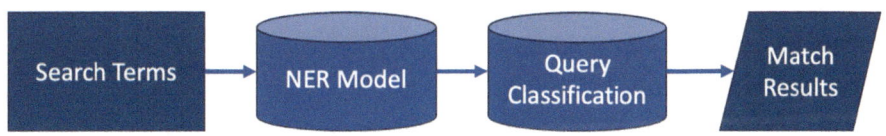

Figure 4-4. *Steps from search to matched products*

The search term can be written in various ways and a few examples are – "Orange sarees below 200," "Orange saree less than 2k," "Orange color sarees under 2000," "Saree orange color price below 2000," "Orang Sarees for under 2000," "Orange Saree below 2000," "Buy orange saree less than 2000," "Orange Sarees < 2000," "2000 below orange saree," "Looking for orange saree < 2000."

The search can be due to singular vs. plural, sequence variation, abbreviation (2000 vs. 2k), spelling mistakes, using symbol (< for less than), spelling variations (such as sari vs. saree or color vs. colour), synonyms (orange vs. saffron), and many more Also, there may be special characters and other issues with search terms.

10-15% of queries are misspelled [5] in web search, and for ecommerce search, a significant volume of search queries are misspelled. Search engines also consider synonyms to improve search. For example, a search for "sneakers" might also include results for "athletic shoes" or "trainers."

Search terms are processed before feeding into NER and classification models, and the processing steps involved are normalization (converting lowercase and remove special characters), stemming/lemmatization (reducing words to their root form, e.g., running to run or sarees to saree), spelling correction (e.g., orang sarees to orange sarees) and stop word removal (e.g., removing the, an, or for).

In ecommerce search, the processed search queries are considered for normalization. Query normalization involves grouping queries based on their semantic meaning or the target products clicked. This process identifies similarities across these products. Using product and category features, search queries are clustered with K-Means clustering on embedding vectors created for the queries, allowing for more accurate and relevant search results.

The processed query goes through query classification and NER models, and the required information for product filtering is extracted. The query classification model identifies the relevant product category: "Saree." And NER successfully extracts key attributes: price <= 2000 and color="Orange" in this example.

Case Study: Custom NER using Transformer Model

Context: *When customers are using certain keywords or phrases in the search query they are looking for specific products. For example, "Vero Moda Floral Dresses" query has three specific user intents – brand, pattern, and product type or category. Managing business rules or logics, it became complicated to manage. Requirement is to migrate to a Machine Learning based system.*

Objective: *Extract Brand, Color, Category and Price entities from the search queries.*

Approach:

Some of the challenges in ecommerce is ambiguity, AND is brand vs. it could be considered as a stop word, w is another brand as a single letter. Also some special characters or numbers are part of the brand names such as Forever 21.

Based on the last 90 days, the most frequent 100k search queries are extracted. From these queries, the single word or two word queries that are direct brand or category queries are excluded.

For the remaining search queries (~60k), the logic-based automated and manual review-based framework is used to create the label data. For example, in the search query "Biba Green Kurta Set."

Figure 4-5 illustrates how entities are defined to create labeled data for training NER models.

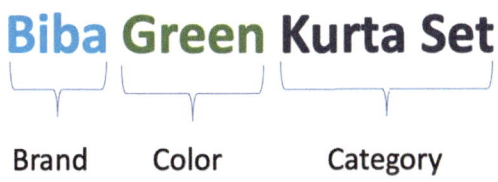

Figure 4-5. *Entities in search query*

The final classes used for the training are S-Brand, I-Brand, S-Color, S-Category, I-Category, I-Price.

Once the labeled dataset is available for training a custom NER, a pre-trained transformer model, ALBERT, was first trained on the catalog dataset so that the model is able to infer the context of ecommerce. The model is then fine-tuned with a cross entropy loss function for the NER labels.

Validation and Evaluation

The model was validated using classification metrics such as recall (the percentage of all relevant entities extracted) and precision (the percentage of identified entities that are correct) on a 20% test sample, with both metrics exceeding 90%. Additionally, daily search queries were normalized, and a sample of 5000 queries was selected after excluding direct searches like brand or category names. As there is no automated method for labeling entities in each query, the model was further evaluated on new data, which included the top 5000 most frequently occurring queries, excluding direct search terms.

The search engine efficiently retrieves all sarees from its vast product catalog that meet these specific criteria. This ensures a targeted selection of products presented to the user, improving the search experience.

Figure 4-6 illustrates the high-level steps involved in using normalized search queries to generate the final list of products for a search term. Buyer queries are normalized before being processed by the query classification model, which helps define the relevant categories for the search, and by the Named Entity Recognition (NER) model to extract relevant entities. These extracted entities are used to apply filter conditions when retrieving products from the catalog. Finally, all matched products are displayed to the users.

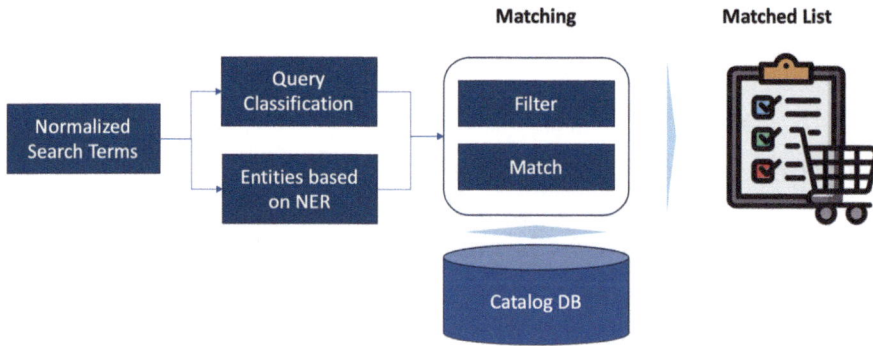

Figure 4-6. *Approach from normalized search terms to linked product list*

Rashmi searches products using search terms such as "Sarees below 2000." When Rashmi gets the search result, search listing pages (SLPs), she reviews the results and decides to search again with a revised search term "printed sarees below 2000" and retrieve a list of sarees that matches the search terms.

Figure 4-7 displays a sample list of products matched to search queries.

Saree Mall
Saree Mall Navy Printed Saree With Unstitched Blouse
₹1149 ₹2899 60% off

Saree Mall
Saree Mall Off-White Printed Saree With Unstitched Blouse
₹1359 ₹3499 61% off

Silk Land
Silk Land Beige & Orange Silk Floral Print Saree With Unstitched Blouse
₹1599 ₹10000 84% off

Kalakari India
Kalakari India Maroon Cotton Printed Saree With Unstitched Blouse
₹1645 ₹3499 53% off

Figure 4-7. *Search results*

In this scenario, in addition to the three steps discussed – query processing, query classification model, and NER model, the search needs text matching of the query term such as "printed" in the product title and description.

Search Algorithms – Text Matching

A heart of ecommerce search is text matching, a method to connect user queries with relevant products in the catalog. The domain of finding relevant products for a given search query is referred to as information retrieval (IR) and it has been studied over decades. The search queries encapsulate customer needs, and typically, the information retrieval fetches a collection of products for a search query in ecommerce.

In the textual or lexical match, the algorithms try to match a search query with products and derive relevance based on the similarity. The similarity is computed as cosine similarity (or similar distance measure) based on frequency of search terms (**TF) and inverse document frequency (IDF)** vectors. The fundamental aspect of text matching, if more

91

often the search terms are present in the product title and description, the higher is its relevance but these terms should also be less frequent across the products.

Figure 4-8 visually illustrates the components of text matching, including how search terms are linked to product details and the role of BM25 in ranking the matched products.

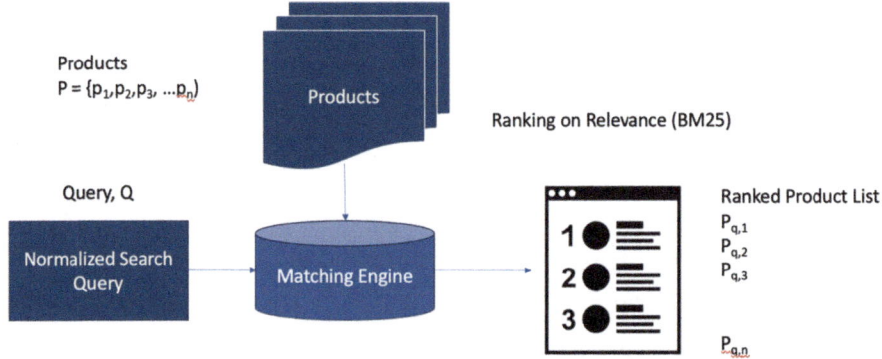

Figure 4-8. *Text Matching - search terms and product details*

Text Matching : BM25 Algorithm

Search queries are matched with product details in ecommerce systems, and many products might contain all the search terms. However, the relevance of these products can vary. For instance, if a search term appears frequently in the query or in multiple product descriptions, this might indicate higher relevance. Conversely, specific or unique terms can help in filtering out less relevant products.

BM25 (Best Matching 25) leverages these concepts to rank products effectively. It considers the frequency of matching terms and their specificity. For example, if a user searches for "high-performance gaming laptop," products with frequent and exact matches for "gaming laptop" will be ranked higher. However, if the term "high-performance" appears only in a few product descriptions, those products will be evaluated differently compared to others where this term appears more frequently.

BM25 calculates the relevance of a product based on both term frequency and document length, adjusting for how often a term appears and how unique it is across the document corpus. This balance helps ensure that products most relevant to the search query are prioritized, improving the search experience for users.

BM25 (Best Match 25) is a common algorithm used for text matching in ecommerce search. This algorithm considers document length and term frequency saturation in addition to term frequency (TF) and Inverse Document Frequency (IDF).

Term Frequency (TF)

Term Frequency measures occurrences of search terms in product details (e.g., product title and description). It is the relative frequency of search terms. For a search term t with a product p is the ratio of frequency of term t with the product p and count of words in product p. It suggests that higher occurrence of search term meaning more relevant is the product for a given search term. There are a few additional formulations for calculating term frequency (TF) [9].

Inverse Document Frequency (IDF)

Some search query terms may be very common and may not indicate any particular relevance for its occurrences in a given product description. Hence, inverse document frequency (IDF) is calculated and it measures how much information the search term provides specificity to a product. In a way, it measures how common or rare it is across all products, it penalizes if it is a common term.

Typically, search platform is based on Apache Lucene and its variants such as Solr and Elasticsearch. The BM25 implementation in Elasticsearch platform is as follows:

$$\sum_i^n IDF(q_i) \frac{f(q_i, D) * (k1 + 1)}{f(q_i, D) + k1 * (1 - b + b * \frac{fieldLen}{avgFieldLen})}$$

q_i is the ith query term,

IDF(q_i) is the inverse document frequency of the ith query term,

f(q_i,D) is count or frequency of ith query term in document D,

fieldLen is length of the field and

avgFieldLen is the average field length across all the products or documents,

k1 and b are the parameters.

Document is equivalent to product details in ecommerce. While text-matching algorithms like BM25 form the backbone of ecommerce search, they can often fall short when users express their needs with imprecise language or use synonyms. BM25 is based on statistical measures and does not have a built-in understanding of the semantic meaning of terms. The BM25 algorithm has k1 and b parameters and selecting optimal values could be a challenge.

BM25 is often preferred over traditional text-matching algorithms like TF-IDF in ecommerce search due to its advanced handling of term saturation and document length variability. Unlike TF-IDF, which can overemphasize term frequency and lacks length normalization, BM25 adjusts for diminishing returns of term frequency and normalizes for document length. So BM25 scores are more reliable in pulling more relevant products.

Search Result Ranking

When searched for "printed sarees below 2000" on an ecommerce platform, it returned over 30,000 products. And in ecommerce, a significantly large percentages of the buyers do not click beyond the first three pages, showing relevant on the top improves the buyer engagement on the ecommerce platform. For showing relevant products on the top could be done based on Popularity or Personalization ranking.

Popularity is mainly based on product performance – how popular are they? How trendy are these?

Product popularity in ecommerce cannot be determined by a single metric like views or purchases alone. For instance, lower-priced products often attract more views, while conversion rates can fluctuate with promotions and discounts. Products displayed prominently on listing pages may receive higher click-through rates (CTR) but could have lower conversion rates. To address these complexities, a composite popularity score is used, combining various metrics to rank products more effectively across ecommerce platforms. This score provides a more comprehensive measure of a product's overall popularity and performance.

A list of product performance attributes along with ranking algorithms can be used for the popularity score. A chapter on the ranking algorithm gives detailed overview on the process to develop ranking models for sorting the products that deliver on customer experience and business performance.

Ecommerce platforms have significant information about the customer preferences especially for the repeat and highly engaged customers. For the customers who have purchased sarees earlier, it should prioritize the products with similar price range, color patterns, and other attributes. Re-ranking of products based on customer preferences is personalization of search results. A detailed approach along with the business impact of personalization in ecommerce is discussed in the personalization chapter.

Search Architecture in Action: From Query to Results

Ecommerce platforms often greet users like José and Rashmi with trending searches upon entering the search bar. This strategy capitalizes on the idea that terms experiencing a surge in popularity are likely to be relevant and interesting to a broader audience. Further, improvements can be made by incorporating geo-location and gender-based personalization for trending searches. This would allow the platform to tailor these suggestions to better reflect the interests of users in specific locations or demographic groups.

Some platforms display a user's previous search terms, assuming they might revisit those searches later. This can be a helpful feature for users who haven't yet made a decision or simply want to pick up where they left off. Also, for some category where repeat purchase for the same category or products are common. For example, on the grocery platform, showing milk or biscuits.

For users entering search queries, the search engine employs spell correction (typically utilizing an ML-based model) to identify and rectify typos in real time and facilitate user experience by retrieving relevant search queries as they type. These suggestions, often called "auto-suggestions" or "type-ahead," are displayed for user convenience.

At Myntra, this approach has demonstrably improved click-through rates (CTR) by 3%, specifically for users who engage with contextual type-ahead suggestions [6].

For search queries like "printed sarees below 2000," the search system undergoes several steps before delivering results. The first step is pre-processing of the query and performing steps like removing special characters, spell corrections, or stemming/lemmatization, etc.

The processed query is fed to NER and query classification models to identify key information from the query. The NER model extracted relevant entities such as brand, color, pattern, etc. Whereas the query classification model gives the category or categories that are relevant to a search term. This includes the product type ("sarees") and relevant attributes like price ("below 2000").

Once product category and other attributes are extracted, the system searches for products with matching attributes in the product catalog. This might involve checking product titles and descriptions. In the final step, the retrieved products are ranked based on factors that indicate their relevance to the user's search.

Ecommerce search results are sorted based on the ranking model that can prioritize either at an overall popularity level or user personalization level.

It is important to give buyer features to select the products from the pulled list of products with the help of a list of relevant filters (facets) that should dynamically adjust for each search query. For instance, when a buyer is searching for "sarees," it might prioritize filters like color and fabric, whereas for "t-shirts," it might prioritize neck type. This tailored approach is facilitated by machine learning (ML) models powering the facet search functionality. Dynamic facets, informed by these models, improve buyer experience by ensuring the displayed filter options are highly relevant and directly applicable to the user's specific search query.

Finally, the buyer is presented with a cohesive search experience, displaying the most relevant products alongside the appropriate filter options at the top of the page.

Figure 4-9 illustrates the key components of the ecommerce search ecosystem. It details the journey from a buyer's search query through the processes of matching and ranking products, and highlights the functionalities available for further refining search results.

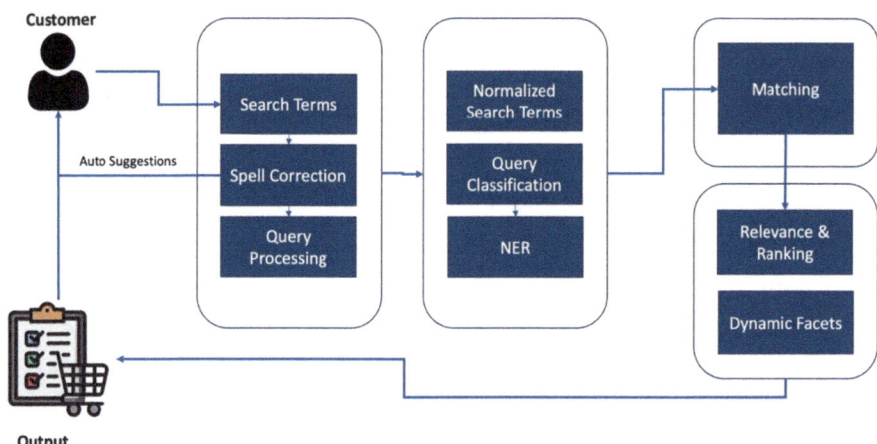

Figure 4-9. *Logical search architecture*

Search functionality in ecommerce is a complex ecosystem involving machine learning models, business rules, evolving product information, and shifting buyer expectations and behaviors. Given these dynamic factors, the performance of search platforms is monitored rigorously. Machine learning models are continuously evaluated and re-trained, with some updates occurring automatically and others following thorough data preparation and deliberation.

New changes are systematically released and tested on a small percentage of users through A/B testing. Once significant improvements are observed, the changes are gradually rolled out to a larger user base.

Semantic Search

Rashmi, while looking for orange sarees, uses terms like "Saffron Color Saree" or "Haldi Color Saree." This highlights a common challenge in ecommerce search, a multiple variation of search queries. On one platform, searching for "Orange Saree" yields 1877 results, while "Saffron Color Saree" shows only 1 result and "Haldi Color Saree" produces no results. This inconsistency demonstrates the challenges in product tagging.

Interestingly, another platform recognizes "Haldi" and displays yellow sarees (1–48 of over 3000 results) when searching for "Haldi Color Saree." Figure 4-10 shows the sample output for the search result.

Figure 4-10. *Semantic search result*

Key challenges in ecommerce search is that the users often use informal language in their search queries, while product titles and descriptions tend to be written in a more formal style. This creates a vocabulary gap that can hinder search accuracy [3]. The techniques like stemming and lemmatization are used to match words with similar roots (e.g., "reading glasses" to "reading glass"), but these can lead to information loss and errors. Similarly, spell-correction methods may not always catch all typos or misspellings in user queries. This can lead to poor search results that don't match the user's intent [7].

Semantic search algorithms incorporate intent and context of the search terms for matching search terms with product catalog. For example, "shoe laces" and "laced shoes" search terms are interpreted by the semantic search engine. Semantic Search architecture handles hypernyms, synonyms, antonyms, morphological variants, and spelling errors in an effective way [7, 8].

Semantic search frameworks rely on dense vectors within an embedding space to capture the contextual meaning of both search queries and products. Deep learning methods convert product information into these vectors, which are then stored for later retrieval [9].

Similarly, user search queries are transformed and stored as vectors as well. Finally, when a user enters a search query, the system links it to its corresponding stored search query vector and compares it against the product vectors to identify matches.

RoBERTa and other transformer models are fine-tuned specifically for the ecommerce context before being employed to create product and search query embeddings. This fine-tuning process involves adapting the models to understand the nuances of ecommerce language and buyer behavior. The embeddings generated from these models are then used to link new buyer search terms to relevant products, enhancing the accuracy and relevance of search results.

Figure 4-11 illustrates the logical steps involved in matching search queries to products using a semantic search framework.

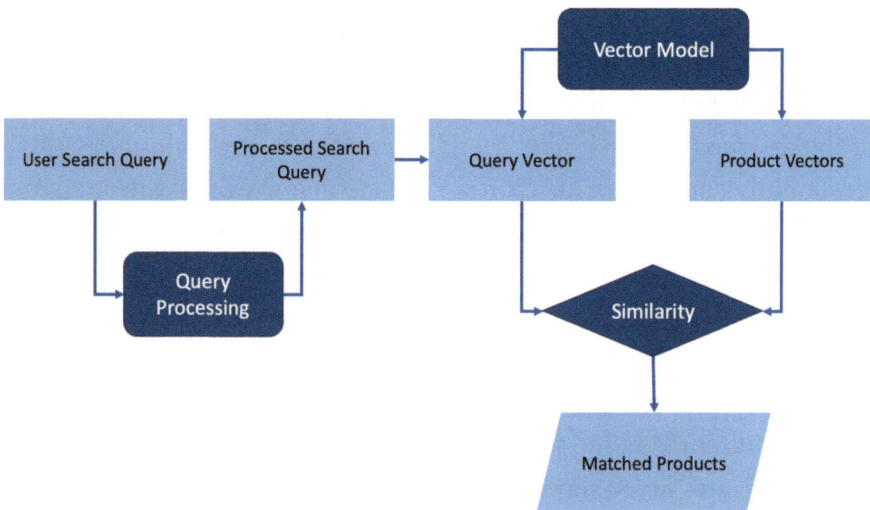

Figure 4-11. *Semantic search steps*

Three main categories of semantic search algorithms [3]:

1. **Factorized Models**: These models independently convert search queries and products into lower-dimensional vector representations (called embeddings) based on their content. These embeddings capture the semantic meaning of the queries and documents or products, allowing for more accurate semantic matching.

2. **Latent Factor Models (LFMs)**: Latent Factor Models (LFMs) use matrix factorization techniques to uncover hidden relationships between words and products (or documents) without directly analyzing their content. By decomposing interaction matrices into latent factors, LFMs can reveal underlying patterns and associations. For instance, techniques like Singular Value Decomposition (SVD) or Non-negative Matrix Factorization (NMF) identify these latent factors, which helps in understanding and predicting buyer preferences and product similarities based on implicit interactions rather than explicit content analysis.

3. **Interaction Models**: Interaction models build interaction matrices between the query and product details (e.g., title and description text). These models capture complex interactions between query terms and product details using deep learning models, leading to more nuanced semantic matching.

Deep Learning for Search Embeddings

Two deep learning approaches are used to create vector representations (embeddings), for both search queries and products are unsupervised and supervised methods.

The recommended supervised approach, detailed in this research paper [3], leverages a transformer-based model called RoBERTa. This process involves two stages:

- **Pre-training:** The RoBERTa model is first trained on a large dataset of fashion-related text (corpus). This initial training equips the model with a strong foundation for understanding language relevant to the search domain.

- **Fine-tuning:** Next, the model undergoes fine-tuning using a technique called triplet loss optimization. This method aims to achieve two goals: reduce the distance between a search query and the product a user clicks on (considered a positive match), and increase the distance between a search query and random products a user doesn't click on (considered negative matches).

Fine-tuning these models requires labeled data to indicate which products are relevant for a search and which are not. Click data is crucial for this process, as it helps identify positive examples (products buyers interacted with) and negative examples (products users did not engage with). Clicks on a product listing after a search indicate a positive match, whereas clicks on irrelevant listings or no clicks at all suggest negative matches. For example, when buyers search for "Red Dress," products that are clicked are considered positive samples, while non-clicked or unrelated products are treated as negative samples. This distinction helps in training the model to better differentiate between relevant and irrelevant products based on user interactions. Additionally, the authors recommended a concept of "narrow" vs. "broad" queries when defining positive and negative samples for training/fine-tuning of the pre-trained model.

Finally, the query and product embeddings learned through this fine-tuned model are used to retrieve relevant products when a buyer enters a search query. The query embedding essentially acts as a reference point for finding products with similar meaning and characteristics within the product catalog.

Conversational Search – Powered by Gen AI

Current search architecture delivers well for the transactional searches and the browse searches. Transactional search involves users who know exactly what they're looking for. For example, a user searching for "printed sarees below 2000" has a clear idea of the product and price range. And when users are exploring a particular product category but haven't yet settled on a specific product, it is called the "browse searches." For instance, a user searching for "sarees" might be browsing for designs and patterns before deciding on the purchase.

The current search architecture might have limitations when it comes to inspirational searches, where customers are actively seeking ideas and inspiration to inform their purchase decisions. For example, a customer searching for "Gift for a toddler birthday" or "Dress for a wedding" is looking for product ideas that fit a specific context. The current search might struggle to deliver on these broader suggestions, potentially limiting itself to exact keyword matches. Ideally, the search engine should recommend products and even across categories to provide a more comprehensive set of inspirational options.

Ecommerce platforms are seeing a surge in "digital window shoppers," similar to their brick-and-mortar counterparts. These customers have the time and inclination to explore the online world, browsing for new trends and product categories. They might be regular visitors to the platform, but their purchases occur less frequently. The current search architecture faces limitations when it comes to inspirational search and engaging digital window shoppers.

The current search architecture may fall short for inspiration-based searches as it is typically limited to a few specific categories. Both text matching and semantic search approaches focus on matching text rather than understanding the context or linking products to their usage or applications. Additionally, product catalogs often lack updated information on when and how to use products or their utility. For instance, catalog descriptions might only include basic details like titles or materials, without indicating that silk sarees are more suitable for special occasions, while cotton prints might be preferred by the older adults and for summer conditions.

Emerging technologies like conversational search offer a promising solution to these challenges. Flipkart, a leading ecommerce in India, launched ChatGPT powered chatbot named Flippi in late 2023 and aims to help buyers who face challenges in making purchase decisions [10]. Flippi helps in finding products that suit their needs. Similarly, Amazon announced Rufus, a new generative AI-powered conversational shopping experience in the first quarter of 2024 [11].

Conversational search allows users to interact with the search engine in a more natural, human-like way, using complete sentences and questions. The conversational search can include context for understanding user intent and preferences, and can facilitate a more open-ended exploration process, helping digital window shoppers discover new trends and products.

Summary

Search serves as a critical funnel for buyer engagement on ecommerce platforms. To create truly engaging buyer journeys, leveraging artificial intelligence (AI) and machine learning (ML) algorithms is fundamental. From processing search queries to linking search terms with products and presenting relevant results, every step is enhanced with AI and ML models to improve outcomes. This augmentation helps boost conversion rates, as more buyers complete purchases through their search journey, and enhances the overall buyer experience.

Entity recognition can identify important entities within search terms and use them for applying filter conditions. For instance, from a search query like "women's running shoes under Rs. 2000," the algorithm might recognize "women" as the target audience, "running shoes" as the product category, and "Rs. 2000" as a price filter.

Search Term Query Classification based on ML algorithms can classify search terms into relevant product categories, helping with efficient matching space and more accurate search results.

Neural Search with Deep Learning: Advanced techniques like neural search, powered by deep learning frameworks, allow the system to go beyond exact keyword matching. It can grasp the semantic meaning behind a search query, enabling it to retrieve products that are conceptually relevant even if they don't contain the exact keywords used by the user.

In essence, AI and ML algorithms are transforming ecommerce search by making it more intelligent, efficient, and user-friendly. This translates into a more engaging buyer journey for online shoppers.

Once buyers are shown the most relevant and/or personalized search results, the buyers explore the detailed products before adding to cart and proceed to purchase. On the journey, the buyers are engaged by showing complementary, similar products, and other products to improve engagement and cart size. These curated choices are powered by a family of algorithms called recommendations. In the next chapter, the focus shifts to understanding the similarity algorithms that drive engaging buyer journeys on ecommerce platforms.

References

[1]   Haixun Wang, The Coming Disruption to E-Commerce Search, Apr-2018, `https://medium.com/gobeyond-ai/the-coming-disruption-in-e-commerce-search-3ffa0eb9a5d9`

[2]   Han Zhang, Songlin Wang, Kang Zhang, Zhiling Tang, Yunjiang Jiang, Yun Xiao, Weipeng Yan, Wen-Yun Yang, Towards Personalized and Semantic Retrieval: An End-to-End Solution for E-commerce Search via Embedding Learning, Jun-2020, `https://arxiv.org/pdf/2006.02282.pdf`

[3]   Sagnik Sarkar, Lakshya Kumar, Neural Search: Learning Query and Product Representations in Fashion E-commerce, Jul-2021, `https://sigir-ecom.github.io/ecom21Papers/paper4.pdf`

[4]   Xiang Cheng, Mitchell Bowden, Bhushan Ramesh Bhange,Priyanka Goyal, Thomas Packer, Faizan Javed, An End-to-End Solution for Named Entity Recognition in eCommerce Search, Dec-2020, `https://arxiv.org/pdf/2012.07553.pdf`

[5]   Qing Chen, Mu Li, Ming Zhou, Improving Query Spelling Correction Using Web Search Results, Jun-2007, `https://aclanthology.org/D07-1019.pdf`

[6]   Hoosein, Real-time context based smart type-ahead
        suggestions, Nov-2028, `https://medium.com/myntra-`
        `engineering/real-time-context-based-smart-type-`
        `ahead-suggestions-316ac7a25107`

[7]   Priyanka Nigam, Yiwei Song and others, Semantic
        Product Search, Jul-2019, `https://arxiv.org/`
        `pdf/1907.00937.pdf`

[8]   Po-Sen Huang, Xiaodong He, Jianfeng Gao, Li Deng,
        Alex Acero, Larry Heck, Learning Deep Structured
        Semantic Models for Web Search using Clickthrough
        Data, Nov-2013, `https://www.microsoft.com/en-us/`
        `research/wp-content/uploads/2016/02/cikm2013_`
        `DSSM_fullversion.pdf`

[9]   Bhaskar Mitra, Nick Craswell An Introduction to Neural
        Information Retrieval, 2018, `https://www.microsoft.`
        `com/en-us/research/uploads/prod/2017/06/`
        `fntir2018-neuralir-mitra.pdf`

[10]   Sudeshna Mitra, Ahead Of Festive Season, Flipkart
        Launches ChatGPT-Powered Shopping Assistant
        "Flippi," Oct-2023, `https://inc42.com/buzz/`
        `flipkart-launches-chatgpt-powered-shopping-`
        `assistant-flippi/`

[11]   Rajiv Mehta, Trishul Chilimbi, Amazon announces
        Rufus, a new generative AI-powered conversational
        shopping experience, Feb-2024, `https://www.`
        `aboutamazon.com/news/retail/amazon-rufus`

CHAPTER 5

Curated Choices Using Art and Science of Recommendations

Overview

It is the month of October, and in Bengaluru, India, most schools have completed mid-term examinations. Leks has two kids, and they are looking forward to a short vacation nearby. After online research and discussions, they converged on Pondicherry (southern part of India and not very far from Bangalore or Chennai) for the trip.

Pondicherry was a French colonial settlement in India until 1954, and now a Union Territory town bounded by the southeastern Tamil Nadu state. The mustard-colored colonial villas and seaside promenade running along the Bay of Bengal are some of the exciting parts. They have heard a lot about this place, and all were looking forward to their trip.

Leks and his friends assembled for a dinner, post the mid-term examinations and a few days before the travel date. All started sharing their plans for the vacations and it was his turn. Leks was impressed with the suggestions that poured in from his friends on places to visit in Pondicherry, eateries, places to go, shopping destinations, and when to do what.

If Leks uses a travel portal with a recommendation engine for his annual travels, the platform will present curated hotel choices based on his historical preferences and budget. Additionally, a generative AI-enabled chatbot can assist with itinerary planning, suggesting activities for each day of the trip. Despite these advancements, many recommendations remain fragmented. Currently, few platforms offer comprehensive end-to-end travel recommendations, covering trip planning, accommodation, restaurant suggestions, and activities.

The friends knew his taste, preferences, budget, and much more. When they suggested places to eat, they factored in that Leks and his family are vegetarians. While suggesting places and when to visit them, they factored in kids and their love for adventure activities.

We all love to get suggestions and feedback when we are exploring options or making choices. We aspire to have a friend that knows about us, has context, and helps us in curated choices that align with our interest and preferences.

What if there is a friend that synthesizes all the information shared with it, and for a given context, explores all the options available and creates a curated choice for you. This new friend is a virtual friend powered by a new technology – Machine Learning and Artificial Intelligence. This "virtual friend" helps all of us across ecommerce platforms like Amazon or Myntra and streaming platforms like Netflix. This virtual friend is called a recommendation engine.

Recommendation engines on ecommerce platforms consider past behavior, preferences, and purchase history to create curated choices (e.g., prepare product list or select a few brands, etc.) that a buyer will be interested in for a given context or visit. These engines not only simplify the decision-making process but also enhance buyer experience.

Recommendation Engines for Ecommerce: Engage Buyers with Curated Choices

Manisha is a housewife and an online shopping enthusiast as most females are and she is a regular shopper at a prominent fashion ecommerce platform in India.

She opens the ecommerce mobile application (app) and can see a wide range of widgets on the app. Today she is interested in exploring the "Kurta set" and decides to follow taxonomy-based navigation and starts her shopping journey.

Figure 5-1 illustrates the browsing journey based on product taxonomy.

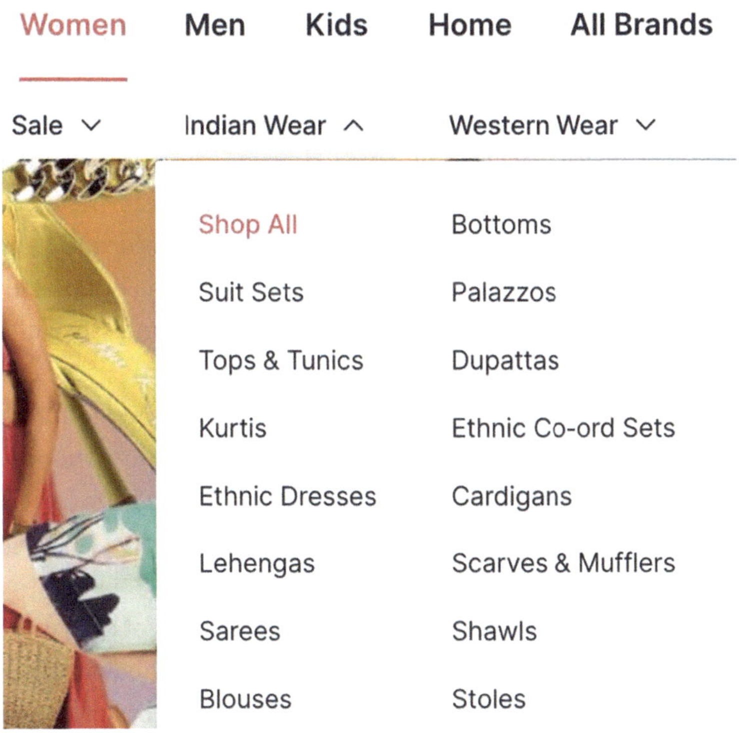

Figure 5-1. *Product taxonomy or hierarchy*

She can see trendy and popular "Kurta Set" on the category page. These are ranked by the popularity-based ranking algorithms, will be discussed in the subsequent chapter and the list is personalized for her history and preferences.

While exploring products, "Similar Products" catches her eye, inviting her to explore further. With each click, the platform presents enticing options that resonate with her style.

When she ventures into detailed product pages, she notices "Customers Who Bought This Also Bought" and that is guiding her through a maze of choices. The result? Her virtual cart brimmed with carefully selected items.

Even on the cart page, she notices "Style On" and entices her to explore products across categories such as Handbag and Sandals and these products matching the kurta-set that she has added to the cart. She adds one pair of sandals to the cart. This is how the platform engages her with options and she loves shopping.

On top of it, every time she returns to the platform, the personalized widgets "Recommended for You" and "Inspired by Your Browsing History" capture her attention and enable her journey on the platform. These unassuming widgets, presenting captivating and enticing options to explore further, offer suggestions that seem to align seamlessly with her tastes and preferences.

Month after month, she finds herself captivated by the platform's suggestions and recommendations, investing both time and money.

Have you ever wondered how ecommerce platforms keep her interested in the platform? How do they know what products to recommend to her?

This is not an isolated story of Manisha but millions of ecommerce buyers worldwide. In India, as well as in other parts of the world, millions of buyers visit ecommerce platforms such as Amazon, Flipkart, and Tmall (a leading Chinese B2C platform with 800 million buyers [3]). These buyers visit the platforms to find products that meet their needs from among hundreds of millions of options available.

Amazon India offers over 168 M products [1] and US Amazon and its marketplace sellers have over 350 million products in the United States [2]. While the vast volume of options available caters to diverse needs, navigating through countless choices can be time-consuming and frustrating. Taxonomy-based navigation may offer some help, but it can also be limited in addressing specific preferences. In such cases, search functionality can come to the rescue by allowing users to quickly find what they're looking for based on their specific criteria.

Leaving users to search and navigate on their own can be costly for ecommerce platforms, as it can lead to poor engagement and lower conversion rates. "The faster we can assist a user in finding the right product, higher are the chances of user conversion" [11]. Creating curated and personalized choices using recommendation-engine that provides a mechanism to get what is relevant to the user, improving buyer experience, engagement, and ultimately conversions.

Recommendation Engine – Business Impact

A well-designed recommendation engine can significantly enhance the buyer experience on an ecommerce platform. By providing curated product choices tailored to each buyer's preferences and browsing history, a recommendation engine can increase buyer engagement, improve conversion rates, and potentially lead to higher transactions per order. Additionally, personalized email or WhatsApp campaigns based on recommendation engine insights can attract buyers to the platform and foster loyalty. Common recommendation algorithms include collaborative filtering, content-based filtering, and hybrid algorithms.

Recommendation Engines are used across industries from healthcare to hotels, in addition to ecommerce [4]. Recommendation engines are utilized across various industries for diverse use cases. In healthcare, they offer personalized treatment recommendations based on medical history, symptoms, and other relevant information, and suggest candidate drugs based on molecular structures and interactions. In the hotel industry, recommendation engines suggest hotels based on travel history and guest preferences, and identify up-sell opportunities by recommending additional services and amenities. Platforms like Spotify and Netflix have harnessed recommendation engines at scale to drive differentiation and enhance user experience, setting them apart in their respective markets.

According to a McKinsey report [5], a recommendation engine contributes over 35% of purchases for Amazon. When buyers are engaged with personalized recommendations on the platform, they have 70% higher conversion rate and increased average order values [5].

Research [5] indicates that when buyers encounter poor recommendations, 38% of US digital shoppers will stop shopping. For a large European IP provider Television (IPTV), up to 30% of the recommendations are followed by a purchase, with an estimated lift factor (increase in sales) of 15% [7].

Recommendations play a critical role in ecommerce from engaging buyers on the platform to increasing conversion rate [6]. Hence, most of the leading platforms have recommendation functionalities ranging from showing personalized products to suggesting complementary items.

A recommendation engine is critical for creating a personalized digital store for each buyer. Jeff Bezos envisioned this concept in early 2000, stating, "If we have 4.5 million buyers, we shouldn't have one store, we should have 4.5 million stores" [7].

Amazon's personalized digital store exemplifies the power of recommendation engines in action [9]. By leveraging a sophisticated blend of advanced machine learning techniques, Amazon creates a nearly individualized digital storefront for each buyer, whether they are on the website or mobile app. When an existing buyer logs in, the homepage is tailored with recommendations based on previous purchases, prominently featuring items the buyer may want to repurchase. Amazon's system also takes into account a variety of factors such as purchase history, day of the week, location, and other contextual data. This comprehensive approach integrates various algorithms, from collaborative filtering to deep learning-based recommendations, ensuring that buyers receive highly relevant and personalized suggestions.

How do algorithms create a personalized store for each buyer? What is the science behind engaging Manisha across her touchpoints on the ecommerce platform?

The Science of Similarity: Crafting Personalized Choices

In crafting personalized recommendations, the pivotal question is, how do these algorithms enable the diverse use-cases that shape our online experiences?

At its core, the primary focus is to meticulously curate choices that are deeply relevant to each individual buyer and the fundamental principle is similarity. Hence, Amit Sharma, a researcher at Microsoft, says on Quora, "recommendation systems are nothing but "similarity hunters." The higher the similarity between a buyer preference and a particular choice, the greater the probability that the buyer will find it appealing and engaging.

There is a long list of recommendation scenarios or functionalities on ecommerce, a few of these are typically referred to below.

- **Similar products**: Finding choices similar to the one the buyer is currently viewing, enhancing their choices within the same category.

- **Frequently bought together**: Recommending products that are often purchased together. For example, if a buyer adds a smartphone to their cart, it might be a screen protector and a case relevant.

- **Customers Who Bought This Also Bought**: Guiding users by showcasing products that were purchased together by other buyers who have also bought this product.

- **Style on**: Encouraging users to explore additional items by suggesting complementary products. For example, if a buyer is exploring a black dress, it might recommend a black blazer to complete the look.

- **Recommended for you**: Personalizing product recommendations based on the buyer's purchase history. It is designed to help buyers discover new products that they might be interested in.

- **Inspired by your browsing history**: This type of recommendation is like the "recommended for you" recommendation, but it is more specific to the products that the buyer has browsed recently. It is designed to help buyers find products that they are already considering.

Consider the scenario of **Trending Items**: When a product gains momentum, indicating that a significant number of buyers are either purchasing or showing interest, it becomes a beacon of relevance. This surge in popularity implies that a buyer who aligns with the preferences of the broader audience, enhancing the likelihood of liking the product. Particularly valuable in the absence of specific buyer insights, the trending items serve as a go-to option, tapping into the collective pulse of the market.

Similar Products, another intricate facet, comes into play when a buyer exhibits interest in a specific product. By showcasing items akin to the one currently being explored, the chances of the buyer finding something they like goes up. The construction of product similarity is nuanced, considering not only explicit data such as historical purchases and browsing but also implicit feedback like ratings and reviews. The recommendation engine meticulously evaluates these parameters, ensuring that the recommended products resonate with the buyer's preferences.

Figure 5-2 illustrates a simple layout for displaying similar product recommendations on a detailed product page of an ecommerce platform.

Figure 5-2. *Similar product widget*

Delving further into the realm of recommendation, we encounter the concept of **Customers Who Bought This Also Bought**. Here, the goal is to unearth new products that might captivate the buyer's interest. The key lies in identifying buyers with analogous tastes and preferences. By examining

117

the choices made by these similar buyers, the engine extrapolates potential interests of the current buyer. The underlying assumption here is profound: similar buyers tend to appreciate comparable choices. By aligning the current buyer's preferences with those of their counterparts, the engine expands the horizon of possibilities, presenting choices that hold the promise of uncharted fascination.

In essence, the recommendation engine's contribution lies in its ability in using buyer behaviors, drawing threads of similarity to create a personalized and enchanting choice list for great experience on eCommerce platform. These algorithms, grounded in the **science of similarity**, redefine how we navigate across ecommerce platforms, ensuring that every choice presented is not just an option but a tailored invitation, designed exclusively for the individual buyers.

Drawing upon the insights gleaned from the science of similarity, we embark on a journey to explore how the interplay of key components orchestrates the delivery of exceptional choices for each buyer.

Recommendation Engine Architecture: Crafting Personalized Choices

In the realm of recommendation engines, data is the oil, serving as the fuel that powers these systems. Algorithms play a crucial role in weaving this data together, using various similarity measures to maximize their effectiveness. Creating effective recommender systems goes beyond simply throwing data at algorithms. Evaluation of its performance and continuously refining it is a fundamental aspect of machine learning-based recommendation systems. Let's delve deeper into the key components that shape the process of creating personalized choices in recommendation engines. These components are

1. Data

2. Similarity measures

3. Algorithms

4. Evaluation

Figure 5-3 for showing Recommendation Engine Architecture and key components

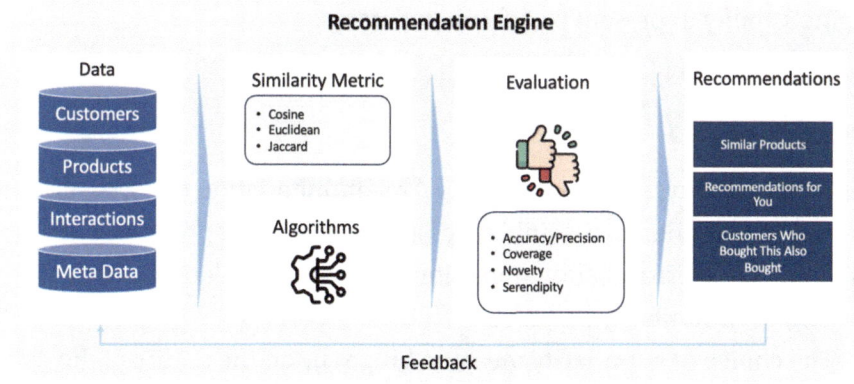

Figure 5-3. *Conceptual facets of recommendation engine*

1. Data

For ecommerce scenarios, an array of data points such as clicks, searches, views, purchases, and cart additions serve as the bedrock for understanding buyer interests and behaviors. These implicit cues shed light on specific buyer preferences, guiding the algorithm toward tailored choices. Some platforms even capture demographic and geographic attributes, providing a nuanced understanding of buyer similarities. Additionally, Ecommerce buyers also provide explicit feedback by providing numerical ratings (such as star ratings) and written reviews for products they have purchased. This feedback can be used to measure whether a buyer liked a particular product or choice. The interactions of buyers and products – purchase/clicks and/or rating are crucial for showing relevant choices to the buyers and these are characterized as implicit and explicit, respectively.

The ecommerce catalog itself stands as a treasure trove of information. Product details encompassing categories, descriptions, prices, and specifications, the catalog enriches the recommendation process. Specifically, it bolsters constructs like "similar products," enhancing the engine's ability to present closely related items.

2. Similarity Measures

Similarity measures are the linchpin of recommendation engines, allowing for the identification of resemblances among buyers or products based on their characteristics and interactions. A recommendation engine is nothing but a similarity machine.

The choice of a similarity measure hinges upon the nature of the data and a recommendation construct. Tailoring the similarity measure to the context ensures that calculated similarities reflect the relationships across key characteristics.

The choice of similarity measure can significantly impact the quality of recommendations. For instance, cosine similarity measures the angle between two vectors and is well-suited for sparse data, such as user-item rating matrices. In contrast, Euclidean distance calculates the straight-line distance between two data points and is more relevant for dense data where vectors have non-zero values. It is also sensitive to the magnitude of the data. In an ecommerce scenario, when dealing with numerical data like product ratings, Euclidean distance might be more effective. However, for collaborative filtering based on sparse product-buyer interaction matrices, cosine similarity often yields better results due to its ability to handle sparsity efficiently.

Here are some common similarity measures:

- **Cosine Similarity**: Ideal for comparing products or buyers represented as vectors, cosine similarity measures the angle between two non-zero vectors in an n-dimensional space. It illuminates the similarity in their directions, offering insights into their compatibility.

- **Pearson Correlation Coefficient**: Suited for handling different rating scales and missing values, this measure gauges the linear correlation between two sets of data points. It illuminates the strength and direction of the relationship, aiding in understanding user-product affinity.

- **Jaccard Similarity**: Tailored for binary data, such as user-item interactions, Jaccard similarity calculates the intersection over union of non-zero elements in two sets. It quantifies the overlap between items, indicating shared user preferences.

- **Euclidean Distance**: Calculating the straight-line distance between two points in an n-dimensional space, Euclidean distance quantifies the dissimilarity between items or users represented as coordinates. It discerns the spatial separation, informing about the dissimilarity of preferences.

- **Manhattan Distance**: Unlike Euclidean distance, Manhattan distance measures grid-based distances, summing the absolute differences of coordinates. This measure is less sensitive to outliers and provides a robust understanding of proximity in a grid, offering insights into buyer-item relationships.

Incorporating these measures judiciously, based on the unique characteristics of the data and the specific recommendation context, ensures that the machine learning algorithms orchestrate a symphony of personalized choices, enhancing user satisfaction and engagement on the ecommerce platform.

Why is Cosine Similarity used for Netflix Recommendation Engine?

Netflix uses cosine similarity to measure the similarity between users and movies. Each user and movie are represented as a vector, the dimensions of the movie vector could include the genres of the movie, the cast of the movie, and the director of the movie. Often a user rates or reviews only a handful of books, leaving most books unrated and similarly, many books receive ratings from only a small subset of users. Cosine similarity is ideal for sparse data because it measures the angle between vectors rather than their magnitudes. It focuses on the non-zero dimensions, making it suitable for high-dimensional, sparsely populated data like user-movie interactions. Also, cosine similarity measure is computational efficient.

3. Algorithms

Recommendation engine algorithms use data and similarity measures to generate personalized recommendations. There are many different recommendation engine algorithms, each with its own advantages and disadvantages. Two foundational approaches are collaborative filtering and content-based filtering. Before delving into the workings of collaborative filtering and content-based filtering, it's important to understand their evolution in the context of recommendation systems.

A Brief History on Filtering Algorithms [8]

Researchers at Xerox PARC (Palo Alto Research Center) developed Tapestry in the mid-1990s, considered to be the first computerized collaborative filtering system. Tapestry enabled hundreds of researchers to annotate news articles with keywords or tags. It also made it easier for researchers to collaborate by detecting users with similar annotation patterns, indicating shared preferences. Leveraging these collaborative patterns, Tapestry filtered and recommended online content to fellow researchers [8, 15].

The collaborative nature of Tapestry, where users' interactions, annotations, and preferences are collectively leveraged to filter and recommend content, gave rise to the term "collaborative filtering."

GroupLens, a research project at the University of Minnesota, made a significant contribution to making collaborative filtering mainstream. GroupLens was one of the first systems to demonstrate the effectiveness of collaborative filtering for recommending a variety of items, including news articles, movies, and music.

GroupLens provided open source software libraries that made it easy for developers to build their own collaborative filtering systems. This helped to accelerate the adoption of collaborative filtering technology.

Amazon.com was a pioneer in using collaborative filtering algorithms to recommend products to its buyers in 1998 [9]. Now recommendation engines are ubiquitous, Alibaba, Amazon, Booking.com, Facebook, Quora, LinkedIn, Instagram, Netflix, YouTube, Pinterest, Spotify, TikTok, Stitch Fix – all of them using recommendation engines to filter and show relevant choices to their buyers [8].

Collaborative filtering and content-based filtering are two of the most used recommendation algorithms in ecommerce. They can be used to generate personalized product recommendations for the buyers. Given the limitations of both content-based and collaborative filtering algorithms, hybrid approaches are often adopted, taking into account the context and various other considerations.

Figure 5-4 illustrates various recommendation algorithms, including collaborative filtering, content-based filtering, and hybrid approaches.

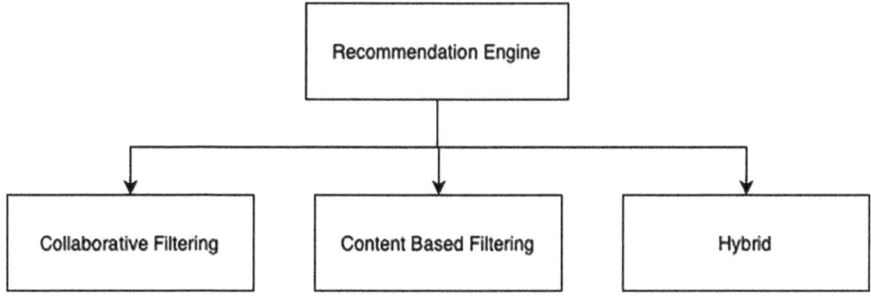

Figure 5-4. *Recommendation algorithms*

Each user/item intersection – each matrix pairing – reveals the degree of user preference for that item. For content-based and collaborative filtering recommenders alike, the utility matrix is where the search for similarities pays off. "Content-based and collaborative filtering recommenders draw upon different types of similarity to populate their respective utility matrixes."

Product/ Customer	Item 1	Item 2	Item 3	Item 4	Item 5
Customer 1		Y			Y
Customer 2	Y		Y		Y
Customer 3		Y		Y	
Customer 4		Y	Y	Y	
Customer 5	Y		Y	Y	Y
Customer 6	Y	Y			Y

One way to conceptualize recommendation algorithms is to view them as predicting the empty cells of a utility matrix based on available information. In the context of ecommerce, this means predicting whether a buyer will purchase or click on a product, given their past purchase or click history.

The machine learning or AI algorithms at Amazon predict a buyer's behavior with 5% accuracy, meaning that the buyer clicks on one of the 20 products shown. This level of accuracy is remarkable considering the vast number of products available on Amazon [12].

Collaborative filtering (CF)

A collaborative filtering is a type of recommendation algorithm that uses choices and preferences of other similar buyers to generate personalized product recommendations for a particular buyer in an ecommerce setting. Collaborative filtering's core assumption is that the buyers with similar preferences buy similar items or rate items similarly. First step in Collaborative filtering algorithms is to identify buyers who have similar preferences to the current buyer and then find the products that these buyers have shown interest in (e.g., purchased or rated).

This diagram shows how a collaborative filtering algorithm works to generate personalized product recommendations for a buyer.

Figure 5-5 for showing working of collaborating filtering, finding buyers with similar tastes and get some products that they explored

Figure 5-5. *Logical view of collaborative filtering approach*

Commonly used similarity measures for identifying similar buyers are Pearson correlation coefficient and the cosine similarity. And buyer interaction data (e.g., purchase and browse history) is a rich source of buyer behavior and interest in ecommerce; hence typically the interaction data is used as input for the similarity calculations.

Collaborative filtering algorithms are very effective at generating recommendations for ecommerce buyers with a lot of interaction data. However, they can be less effective for buyers who are familiar with the system (cold start problem) or have unique preferences.

Content-Based Filtering

Content-based filtering is a type of recommendation algorithm that generates personalized product recommendations for a particular buyer by identifying features of the products that the buyer has purchased in the past and then recommending similar products that share those features.

Content-based recommenders need to create both product profiles and a buyer profile. An item profile is a matrix of "items to features" describing key item attributes such as brand, category, color, style, material, etc. in ecommerce. Buyer profiles rely on the same features as the constructed item profiles to find liking or preferences of the buyers toward each of these features. For example, a buyer likes Adidas Brand, Men Category, and Blue Color.

This diagram shows how a content-based filtering algorithm leverages historical interactions data to generate personalized product recommendations for a buyer.

Figure 5-6 for showing the working of content-based filtering, finding products similar to buyer purchased products

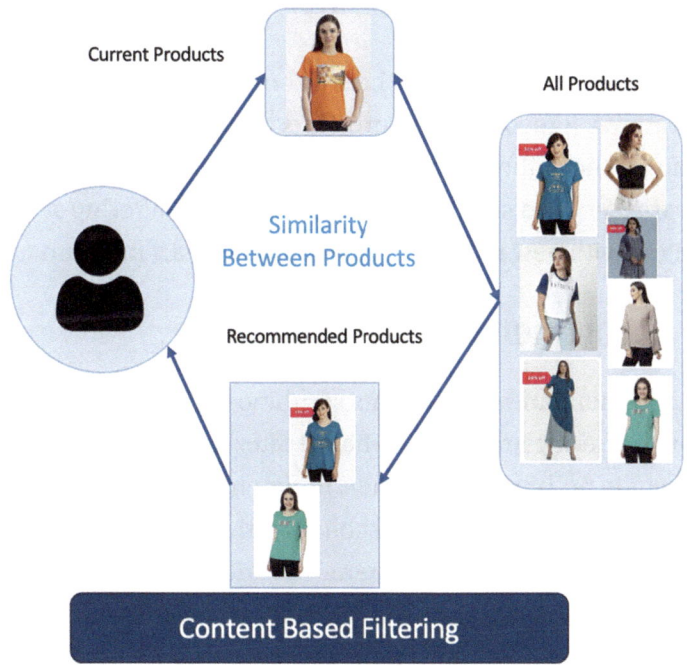

Figure 5-6. *Logical view of content based filtering approach*

For example, if a buyer has purchased running shoes at a price range of Rs. 2000 to 3000, a content-based filtering algorithm would recommend other running shoes in the same price range.

Common features used in content-based filtering include brand, price, color, style, and category. However, any feature that can be used to describe a product can be used in content-based filtering.

Spotify "model the behavior of every single user on Spotify – their tastes, based primarily on their listening habits, what features they use on Spotify and also what artists they follow" to create a weekly playlist for each of the users [8].

Content-based filtering is a powerful algorithm and can be used for generating personalized product recommendations; ecommerce companies can help buyers by finding products that they are most likely to purchase or click.

Content-based filtering relies heavily on buyers' past preferences and product features. It recommends products like the ones already purchased; hence limiting the ability to recommend new products that the buyer may not be aware of (called serendipity). Content-based filtering struggles when dealing with new buyers or products; cold start problem.

Hybrid Recommendation Algorithms

Hybrid recommendation algorithms combine the strengths of collaborative filtering and content-based filtering to recommend products to users that are both relevant to their preferences and new to them.

Combining content-based and collaborative filtering algorithms can enhance recommendation systems in various ways. One approach involves blending the recommendations from both methods by assigning weights, such as allocating 40% of the recommendations to collaborative filtering and the remaining 60% to content-based filtering. Another method involves scoring products based on weighted scores from both algorithms and presenting the final recommendations accordingly.

Contextual factors also play a crucial role in determining which approach is more effective. For instance, when limited information is available about buyers, content-based recommendations may be prioritized. Conversely, for highly engaged users with rich interaction data, collaborative filtering tends to deliver more relevant recommendations.

For the new buyers on the ecommerce platform, hybrid algorithms leverage content-based recommendations to address the cold start problem. Hybrid systems also incorporate demographic information, user context, and historical data to refine recommendations.

Similar user-based product recommendations can help in addressing the issue of lack of variety in content-based filtering algorithms.

Here are a few examples on how hybrid algorithms are used by the ecommerce platforms. When a buyer visits the homepage of an ecommerce website, a hybrid recommendation algorithm can recommend products to them based on their past purchase and browsing history, as well as the products that other buyers with similar preferences have purchased.

Also, when a buyer views a detailed product page, a hybrid recommendation algorithm can recommend other products that the buyer may be interested in, such as complementary products or products from the same brand or category.

Hybrid recommendation algorithms offer benefits over traditional recommendation algorithms, including

- **Improved accuracy**: Hybrid recommendation algorithms are typically more accurate than traditional recommendation algorithms because they consider a wider range of factors.

- **Better serendipity**: Hybrid recommendation algorithms are better at recommending products that are new to the buyer and that they may not have been aware of on their own.

- **Reduced cold start problem**: Hybrid recommendation algorithms are less affected by the cold start problem than traditional recommendation algorithms because they can use content-based filtering to recommend products to new users even if they have not interacted with the system much yet.

4. Evaluation

E-commerce platforms like Amazon manage millions of products and rely on recommendation engines to deliver highly relevant suggestions to buyers every day. How do you evaluate if the recommendation system is doing its job of meeting buyer preferences?

Ecommerce platforms like Amazon or Myntra show personalized product recommendations to millions of their buyers. Evaluating the performance of these recommendation systems is a complex task, as the focus and priorities may differ across organizations. However, there are few key metrics that can be used, such as accuracy, coverage, and novelty. If the objective is to increase user satisfaction, then metrics such as buyer engagement (such as clicks and time spent, etc.) and satisfaction may be more important than accuracy.

Before any recommendation algorithm goes live on an ecommerce platform, the output is evaluated and some of the evaluation metrics considered are

- **Accuracy:** Measures how well the recommendation engine predicts the relevance of the products to the buyers.

- **Coverage:** Measures proportion of the products getting recommended to the buyers.

- **Novelty:** Measures the proportion of recommended products that are new to the buyers.

In a commercial setting, business metrics are significantly important. Some of the business metrics influenced by the recommendation engine are

- **Click-through rate (CTR):** When products are recommended to the buyers, how many buyers like and click them. And click through rate helps in measuring that and it is the percentage of buyers who click on a recommended product.

- **Conversion rate (CVR):** The goal for ecommerce platforms is to sell the products to their buyers. That means buyers buy the products that are recommended to them. Conversion rate captures proportion of buyers who end up purchasing in each of the visits on the ecommerce platform. Conversion rate is the percentage of buyers who click on a recommended product and then purchase it.

- **Purchase revenue:** Typically, lower per unit price products have higher conversion rate. The ecommerce platform also optimizes the recommendations to increase overall revenue from the buyer visits. Hence, the total amount of revenue generated from recommended products is an important consideration of the evaluating recommendation engine.

Zelano, a fashion ecommerce platform, "Measuring click-through rates is a core metric for recommendations, but you cannot forget about tracking gross revenue" [14].

Ecommerce platforms like Amazon or Myntra use A/B testing to evaluate the performance of different recommendation algorithms and approaches. In A/B testing, it randomly assigns buyers to different groups and then shows products based on different recommendations.

131

Performance on these metrics is compared for each group to determine which recommendation algorithm or approach performs the best. For proper A/B testing, each of the groups need to have a large enough buyer base to get statistically significant results.

In 2021, Amazon generated 35% of its sales from recommended products. This means that Amazon's recommendation engine is responsible for billions of dollars in revenue each year, and for helping buyers to discover new products that they may not have been aware of [5].

Now, a few scenario use-cases are considered and then above recommendation architecture components are explained with the steps.

Fashion Ecommerce: Recommendations for You

Manisha is a regular shopper to fashion ecommerce platforms. When she visits the platform, she is greeted with a curated list of products under the "Recommendations for you" widget. One of the approaches to power this widget is collaborative filtering algorithm. Collaborative Filtering (CF) is the most successful recommendation technique to date [10].

Fashion ecommerce has an interesting business model, the trends evolve at a rapid pace, more products are added each day and there is a huge volume of products for any given category.

Considering each of the products will be purchased by few buyers only, buyer product purchase interaction data is hugely sparse. So, one approach could be to leverage buyer interest and affinity data. We can create handcrafted features that capture buyer interest or indicative of buyer preferences and then collaborative filtering algorithms can be leveraged.

Handcrafted features, curated by domain experts or based on the taxonomy, capture nuanced buyer preferences. For example, in fashion ecommerce, the features can capture brand affinity, category preferences, price ranges, size preferences, color preferences, style attributes, and many more. Additionally, the features include demographic data (gender, location), and device-related information (app vs. web, iOS vs. Android) related characteristics. The handcrafted features help to identify buyer style preferences, which are often difficult to capture using traditional interaction (purchase or ratings) data.

User-Based Collaborative Filtering (UBCF) coupled with handcrafted features offers a strategic approach to provide tailored recommendations and address buyer preferences effectively. A well-defined user profile can differentiate a more personalized or customized recommendation system from a conventional system [14].

In user-based collaborative filtering, the recommendation process involves comparing users' preferences and behaviors. When handcrafted features are involved, creating a matrix involves representing users and their features in a structured way. We construct a user-feature matrix where each row corresponds to a buyer and each column for a feature.

The values in the matrix indicate the strength or relevance of a particular handcrafted feature for a specific user. In a simplified scenario, it can be a binary (0 or 1) for the presence or absence of a feature.

In the first step, for the millions of the buyers on the fashion ecommerce platform, the handcrafted features, including brand affinity, category preferences, price ranges, gender, age, and geographic location, are created. These features are transformed into a binary user-feature matrix, where 1 indicates a purchase and 0 indicates no purchase for the user and feature combinations.

Once the user–feature matrix is constructed, the next step in collaborative filtering is to calculate the similarity between users. This is done to identify the most similar buyers based on handcrafted features. And the two common similarity metrics used for this purpose are Jaccard similarity and cosine similarity. Nearest neighbor algorithms are then used to find the top N most similar buyers to a given user. These algorithms identify the users whose preferences are most similar to the target user's preferences.

Figure 5-7 illustrates how collaborative filtering algorithms compare buyer characteristics to identify similar buyers and generate recommendations.

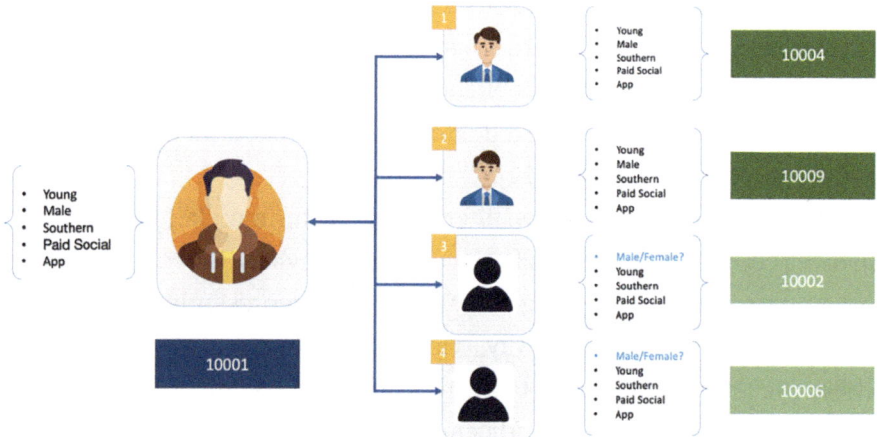

Figure 5-7. *Similar buyer identification*

Based on the nearest neighbors, the top N similar buyers are selected. For these buyers, all products preferred by them are selected as the consideration set. The underlying assumption is that if these buyers are similar in taste and preferences to the user under consideration, the products preferred by them will also be of interest.

For the final curated recommendations, the products that are already purchased by the user are excluded. This is especially important in the fashion category, where buyer may not want to purchase the same product again.

This approach helps in finding curated fashion products that have been explored and liked by similar buyers, providing more personalized and relevant recommendations for the buyer/user.

Figure 5-8 displays products favored by similar buyers, which can be included in the final recommendation set.

Figure 5-8. *Recommended product listing*

Recommendation systems based on Collaborative Filtering techniques have successfully worked for many scenarios [14]. When certain users exhibit similar characteristics due to hand-crafted features, adopting User-Based Collaborative Filtering for personalized recommendations becomes easier. By preparing a user matrix with handcrafted features, sparsity in the data can be reduced. Additionally, user–user relationships are leveraged for recommendations, it brings novelty into the choices.

However, there are some challenges associated with this approach. One such challenge is overfitting, especially when there are limited user profile features and interaction data in ecommerce settings. Another challenge is the computational expense of calculating the similarity matrix between all pairs of buyers, which can be prohibitive for large datasets.

The functional and algorithmic foundations of recommendation engines have been extensively detailed. However, practical implementation extends beyond the framework and algorithms, as recommendations in ecommerce platforms or similar digital ecosystems are often delivered in near real time. The complexity involved in generating these recommendations necessitates advanced technological innovations and robust ecosystems.

Generating recommendations for millions of users while considering billions of data points requires a system designed for massive scalability. While the book focuses on functional and algorithmic aspects, understanding high-level technology considerations is also crucial.

Cloud technology plays a pivotal role in enabling recommendations at scale. The architecture choices vary depending on the use case. For instance, similar product recommendations might not require full computation in real time, although recommendations are served instantaneously. Such systems often apply business rules and product availability checks before presenting options to users.

In contrast, real-time recommendation engines rely on live click data, requiring real-time data streaming and significant computational resources. The architecture for these systems involves more complex technology components to handle the higher demands of real-time processing and scalability.

Similar Products – Personalized Product Recommendations

Similar product recommendations are one of the most common features across ecommerce platforms. When buyers reach the product detail page (PDP), there are typically two ways to reach similar product recommendations: one option is over the product image, and the second a widget lower in the page.

Figure 5-9 illustrates an example of how similar product recommendations are presented on an ecommerce platform.

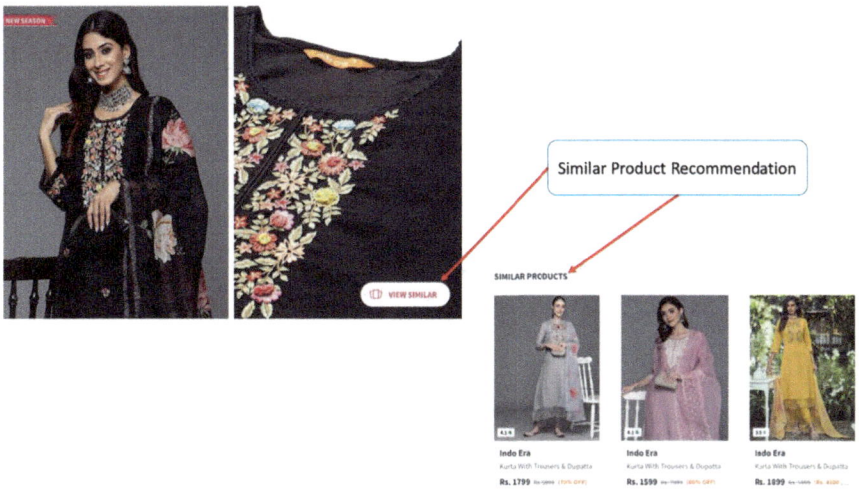

Figure 5-9. *Similar Products*

Around 20–30% buyers visit product-detailed pages on a fashion ecommerce and 15% of these buyers explore similar products. The conversion rate is typically 3–4X for the buyers who use the similar product recommendations, as these buyers are really interested in purchasing the products.

Similar product recommendations can be created using either content-based or collaborative filtering methods. In content-based filtering, product attributes such as color, pattern, fabric, sleeve type, collar type, etc. are used to recommend similar products. However, these features are not fixed, as trends and preferences change over time. Another challenge with content-based filtering is the manual tagging of these attributes [11].

Unlike content-based similarity measures, item–item collaborative filtering algorithms use buyer signals instead of a taxonomy to compute item–item similarity and underlying considerations are co-rating or co-purchases. If products are rated or purchased by the same buyers, these two products have similar latent factors that are relevant to the buyers.

Item-based recommendation techniques (also known as model-based) analyze the user–item matrix to identify relations between the different items, and then use these relations to compute the list of top-N recommendations [16]. Amazon's "Customers Who Bought This Also Bought" feature is also a prime demonstration of Item–Item Collaborative Filtering in action.

At Myntra, a large ecommerce platform in India, item–item collaborative filtering approach is used for creating personalized similar product recommendations [11]. Buyer product interaction data is used for creating a utility matrix – user–item interaction matrix. For user–item interaction, three user signals, i.e., product clicks, addition of products to cart, and checkout are considered and then estimate the relationship between number of product clicks, add to carts, and checkout using a linear classification model [10].

This matrix is highly sparse as each buyer typically interacts (purchase or view) with only a handful of products out of the vast array of options available. This sparsity poses a challenge for traditional recommendation algorithms, which often struggle to handle large and sparse datasets effectively. Matrix factorization methods are used to convert buyers and products interaction matrix to a lower-dimensional latent space. One of the methods used for decomposing the interaction matrix into lower dimensional metrics is Alternating Least Squares (ALS) [21].

The input user–item interaction matrix is transformed into two dense matrices: the Buyer Features Matrix (U) and the Product Features Matrix (V). These matrices capture latent factors or features, with U capturing buyer preferences and V containing product attributes that buyers consider implicitly. By comparing each buyer (row in the buyer features matrix) to products (columns in the product features matrix), we can identify buyer preferences across products.

Figure 5-10 illustrates how the utility matrix is connected to both the Buyer/Customer Feature Matrix and the Product Feature Matrix.

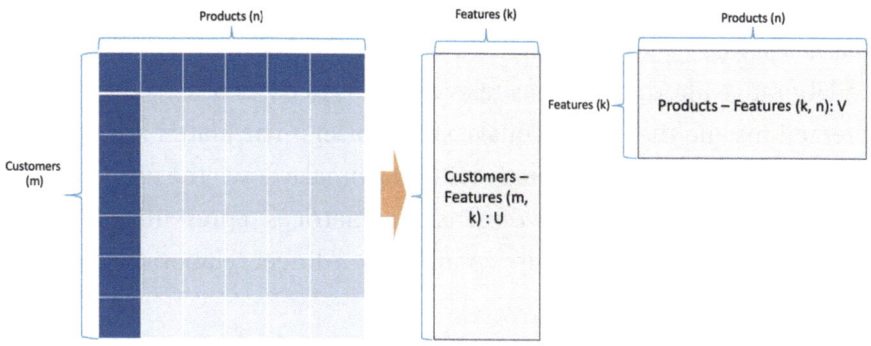

Figure 5-10. *Matrix decomposition*

Using the product feature vectors, a non-personalized candidate is generated by comparing the product under considerations to the rest of catalog products using cosine similarity between the vectors. Then top N similar products were selected as candidates for the similar product recommendations.

For personalized recommendations, the paper used both product–product similarity scores and user–product similarity scores. This approach helps in preserving the context as well as the effect of personalization.

In addition to these fundamental approaches – collaborative filtering and content-based filtering, deep learning frameworks are employed to address some of the challenges around scalability, leveraging diverse information and performance improvements. In the next section, an overview of using deep learning for collaborative filtering and content-based filtering are discussed.

Neural Collaborative Filtering (NCF)

Collaborative filtering suggests relevant items to users based on their past interactions and the interactions of similar users. The Matrix Factorization (MF) method within collaborative filtering breaks down the user–item interaction matrix into lower-dimensional matrices representing latent factors. MF struggles to capture complex, non-linear relationships between users and items.

Neural Collaborative Filtering (NCF) is a powerful approach that uses deep learning to enhance recommender systems. By employing neural networks, NCF can capture intricate, non-linear relationships between users and items [13]. This enables NCF to leverage implicit data such as clicks and purchases, resulting in more accurate and personalized recommendations. The original NCF paper [17] introduced the use of a multi-layer perceptron to learn the user–item interaction function, highlighting its effectiveness in modeling complex user–item relationships.

At a high level, the NCF architecture has three main components – embedding layer, interaction layer, and output or prediction layer. The input item and user vectors are categorical feature vectors, both are concatenated and used as input to the embedding layer. These embeddings represent the latent features of users and items. Interaction layer utilizes a multi-layered neural network (like a multi-layer perceptron) to model user interactions with items. The user and item embeddings are fed into these layers, allowing NCF to model complex, non-linear relationships that go beyond what traditional Matrix Factorization can capture.

The output layer takes the interaction representation generated in the previous layer and predicts the likelihood of a user interacting with an item. This prediction typically involves a score indicating the potential for a user to click on, add to cart, or purchase a particular item.

Figure 5-11 illustrates the functional architecture components of neural collaborative filtering.

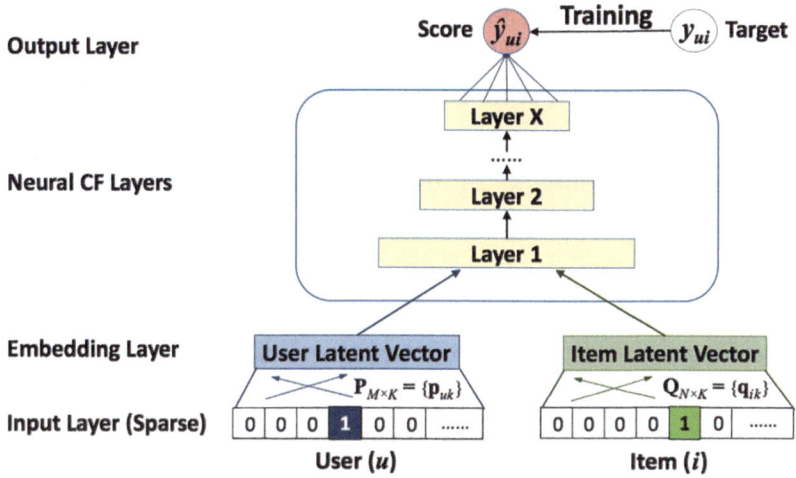

Figure 5-11. *Neural collaborative filtering*

Source: https://arxiv.org/abs/1708.05031

The NCF utilizes an objective function (often pointwise or pairwise loss) to assess its prediction accuracy. By continuously minimizing this loss function, the model learns and refines its parameters. The observed user interactions (clicks, purchases) are compared with the predicted scores to determine how well the model is performing.

Content-Based Filtering Using Deep Learning

Like Neural Collaborative Filtering (NCF), the traditional content-based filtering (CBF) can be augmented by leveraging artificial neural networks and deep learning. The traditional CBF struggles to handle the complexities of textual, video, and audio data. Deep learning-based models have emerged as a transformative solution, enabling the integration of rich item information such as product descriptions, specifications, and acoustic features of music into recommendation systems. This approach not only enhances recommendation accuracy but also eliminates the need for manual feature extraction and data labeling [18], making it a more efficient and scalable solution for real-world CBF applications.

In the realm of ecommerce, product descriptions, reviews, and image data can provide invaluable insights for recommending similar products to buyers. Deep learning-based models have become indispensable tools for effectively processing and utilizing this diverse range of data [19]. A typical deep learning architecture for CBF comprises Feature Extraction Layer (e.g., extracts relevant features from the product content), Embedding Layer (e.g., transforms product categories, brands, and tags, etc. into numerical representations), Interaction Layer (e.g., model interactions using Multi-Layer Perceptron), and Output Layer (e.g., comparing interaction with predicted interaction by binary loss function).

The combination of these components enables deep learning-based CBF models to effectively capture and utilize rich item information, leading to more accurate and personalized recommendations.

Style Up or Complete My Look Using Computer Vision

In the realm of advanced recommendation algorithms, "Style It Up" or "Complete My Look" stands out as a prominent feature. This feature aims to showcase products that seamlessly complement a selected product, effectively completing the user's desired look.

As noted by [20], this functionality requires a sophisticated model that not only identifies the appropriate product types (articles of clothing) to construct a complete outfit but also selects individual items that harmonize stylistically, considering factors like brand, price, color, and overall theme to create a cohesive ensemble.

In an ecommerce setting, for instance, a buyer seeking a complementary T-shirt to match their uploaded trouser image can utilize the Style Up or Complete my Look feature. The image processing component extracts features from the uploaded image, identifying relevant aspects such as color, style, and occasion. These features are then fed into the recommendation model, which generates a list of recommended T-shirts. The buyer can then browse the recommendations and select the one that best suits their preferences.

Overall, Style Up or Complete my Look using Computer Vision emerges as a promising feature capable of generating accurate, personalized, and scalable style recommendations for users.

Conclusion

Recommendation engines, Machine Learning (ML) and Artificial Intelligence-(AI)based methods, are transforming the way we interact with ecommerce platforms, by offering a personalized and efficient shopping experience. Amazon has been a pioneer in leveraging the recommendations at scale in ecommerce. 35% of Amazon purchase transactions are enabled by recommendation systems.

An array of recommendations use-cases across ecommerce platforms are on the display, for example, recommendations for you, frequent purchases together, similar products, style up or complete the look. A successful recommendation system requires a multi-faceted approach that considers data, algorithm, similarity metric, and evaluation methods. Two most common algorithms used in a recommendation engine are collaborative filtering and content-based filtering.

Collaborative filtering leverages past interactions (purchases, ratings, clicks) to identify users with similar tastes and recommend items that those similar users have preferred. This approach introduces new and interesting options to the users, they might not have considered otherwise.

Content-based filtering considers features like product descriptions, genres, categories, and user reviews to recommend items with similar attributes to those a user has interacted with in the past. It creates choices that are similar to what a user has already interacted with, leading to a higher chance of finding these choices relevant.

Deep learning has been a major success across applications and the recommendation systems are not left behind. Both collaborative filtering and content-based filtering approaches are augmented with deep learning frameworks. The Neural Collaborative Filtering (NCF) has shown remarkable improvement in performance and delivers results at scale. Also, deep learning helps expand content-based filtering to be a multi model – leverage a variety of data – from images to interaction data for generating the recommendations.

Convolution Neural Network (CNN) architecture is transforming Style Up or Complete the Look recommendation use-cases.

When a long list of choices is aligned with the user's taste and preferences, how to prioritize the most relevant on the top. Arranging choices in order to improve buyer experience and business outcome is a common challenge across search, recommendations, and category navigation journeys. This challenge is solved using a family algorithm called ranking algorithms. This is the topic for the next chapter.

References

[1]   Emily Dayton, Amazon Statistics You Should Know: Opportunities to Make the Most of America's Top Online Marketplace, `https://www.bigcommerce.com/blog/amazon-statistics/`

[2]   Aayushi Sharawat, Ten Amazon Statistics You Need To Know In 2024, `https://www.shiprocket.in/blog/ten-amazon-statistics/`

[3]   WPIC, Exploring the Top 5 Chinese E-commerce Platforms in 2024, Jul-2024, `https://www.wpic.co/blog/top-5-chinese-ecommerce-platforms/`

[4]   Pegah Malekpour Alamdari, Nima Jafari Navimipour, Mehdi Hosseinzadeh, Ali Asghar Safaei, A Systematic Study on the Recommender Systems in the E-Commerce, Jun-2020, `https://www.researchgate.net/publication/342222975_A_Systematic_Study_on_the_Recommender_Systems_in_the_E-Commerce`

[5]   Ian MacKenzie, Chris Meyer, and Steve Noble, How retailers can keep up with consumers, Oct-2023, `https://www.mckinsey.com/industries/retail/our-insights/how-retailers-can-keep-up-with-consumers`

[6]   Krista Garcia, The Impact of Product Recommendations, Aug-2028, `https://www.emarketer.com/content/the-impact-of-product-recommendations`

[7]   Francesco Ricci, Lior Rokach,Bracha Shapira,Paul B. Kantor, Recommender Systems Handbook, Springer Publication

[8]   MICHAEL SCHRAGE, RECOMMENDATION ENGINES, MIT Press

[9]   Leslie Walker, Amazon Gets Personal With E-Commerce, Nov-1998, `https://www.washingtonpost.com/wp-srv/washtech/daily/nov98/amazon110898.htm`

[10]   Amber Madvariya, Sumit Borar, Discovering Similar
        Products in Fashion E-commerce, https://sigir-
        ecom.weebly.com/uploads/1/0/2/9/102947274/
        paper_19.pdf

[11]   Pankaj Agarwal, Sreekanth Vempati, Sumit Borar,
        Personalizing Similar Product Recommendations in
        Fashion E-commerce,Jun-2028, https://arxiv.org/
        pdf/1806.11371.pdf

[12]   Ajay Agrawal, Joshua Gans, Avi Goldfarb, Prediction
        Machines – The Simple Economics of Artificial
        Intelligence, Harvard Business Review Press

[13]   Marcel Kurovski, Deep learning for recommender
        systems, Feb-2018, https://medium.com/berlin-
        tech-blog/deep-learning-for-recommender-
        systems-48c786a20e1a

[14]   Dr. Mikio Braun, Recommendations Galore: How
        Zalando Tech Makes It Happen, Dec-2026, https://
        engineering.zalando.com/posts/2016/12/
        recommendations-galore-how-zalando-tech-makes-
        it-happen.html

[15]   Samit Chakraborty, Md. Saiful Hoque, and others,
        Fashion Recommendation Systems, Models and
        Methods: A Review, Jul-2021, https://www.mdpi.
        com/2227-9709/8/3/49

[16]   Mukund Deshpande, George Karypis, Item-Based
        Top-N Recommendation Algorithms, Jan-2003,
        https://conservancy.umn.edu/bitstream/
        handle/11299/215545/03-002.pdf

[17]   Xiangnan He, Lizi Liao, Hanwang Zhang, Liqiang Nie,
Xia Hu, Tat-Seng Chua, Neural Collaborative Filtering,
Aug-2017, `https://arxiv.org/abs/1708.05031`

[18]   SHUAI ZHANG, LINA YAO, AIXIN SUN, YI TAY, Deep
Learning based Recommender System: A Survey and
New Perspectives, Jul-2029, `https://arxiv.org/`
`pdf/1707.07435.pdf`

[19]   Narges Yarahmadi Gharaei, Chitra Dadkhah,
Content-based Clothing Recommender System using
Deep Neural Network, Mar-2021, `https://www.`
`researchgate.net/publication/351423084_Content-`
`based_Clothing_Recommender_System_using_Deep_`
`Neural_Network`

[20]   Najmeh Forouzandehmehr, Personalized "Complete
the Look" model, Jan-2023, `https://medium.com/`
`walmartglobaltech/personalized-complete-the-`
`look-model-ea093aba0b73`

[21]   Reza Zadeh, Matrix Completion via Alternating Least
Square(ALS), 2015, `https://stanford.edu/~rezab/`
`classes/cme323/S15/notes/lec14.pdf`

CHAPTER 6

Ranking: Science of Sorting in Ecommerce

Introduction

For a shopping journey on an ecommerce platform, the search feature offers a faster way to find desired products. Buyers who employ search tend to convert or place orders at a higher conversion rate, around four times higher than those who don't. Buyers who use search functionality typically have a clear purchase intent, and search expedites their journey toward finding products. Ranking algorithms further enhances this journey by helping them discover the most relevant products more efficiently.

On the other hand, a considerable buyer cohort, especially in the realm of fashion ecommerce, consists of "explorers." These individuals are actively seeking variety and options that align with their unique preferences. For this cohort, a virtual sales assistant proves indispensable, providing tailored recommendations and suggestions for products they may find appealing. Essentially, the recommendation engine serves as the virtual sales assistant in fashion ecommerce. In the preceding chapter, the process of curating personalized choices for each buyer is deliberated upon. In addition, these curated recommendations are further optimized using ranking algorithms to enhance relevance for buyers, thereby driving key business outcomes such as conversion rates and gross merchandise value (GMV).

© Ramgopal Prajapat 2024
R. Prajapat, *AI-Powered Ecommerce*, https://doi.org/10.1007/979-8-8688-0923-1_6

In this chapter, we delve into the concept of a machine learning twin, known as ranking, often utilized alongside search or recommendation engines. Ranking plays a pivotal role by reordering search results or recommendations, thereby enhancing relevance and ultimately improving conversions and GMV. The role of ranking algorithms extends beyond merely complementing search and recommendations; they have wide-ranging applications in stand-alone scenarios. Ranking algorithms are extensively used to prioritize categories on home pages and even to rearrange payment options on checkout pages. This chapter focuses on the science behind ranking, including the framework for developing ranking models and the processes for creating labeled data and features.

Ranking in Ecommerce

Every day, hundreds of thousands of shoppers visit ecommerce platforms and utilize search functionality to find products. Manisha, a regular ecommerce shopper, recently wanted to explore sarees on her preferred mobile app. She entered the search term "printed saree" and was presented with a message indicating over 77,000 available options.

Home / Clothing / **Printed Sarees**

Printed Sarees - 77323 items

Figure 6-1. *Display of product counts*

With over 77,000 options available and each page displaying only 50 products, it would be overwhelming for Manisha to explore all the choices. This challenge is not unique to her search but is common across many queries, as a significant proportion of ecommerce searches consist of short phrases, typically fewer than four words. Moreover, not all products are equally relevant; some listings may be outdated and irrelevant due to evolving fashion trends, while others are trending and garnering substantial attention from buyers. What mechanism should be used to display the top 50 most relevant products to buyers? The ranking algorithm is one such mechanism.

The ranking system prioritizes and displays the top 40 most relevant products. Details of how this ranking is achieved will be explored further in the chapter.

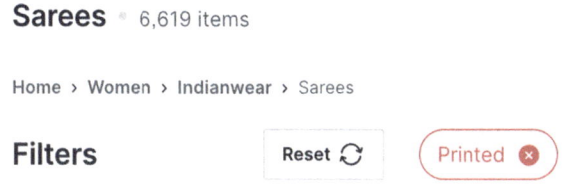

Figure 6-2. *Navigation and product counts*

Manish explores another ecommerce platform and navigates through its product taxonomy. She encounters over 6000 options, but only 40 products are displayed on the first page. These 40 products are carefully selected and arranged to ensure they are the most relevant to her search.

The science of arranging these products for the search or navigation is referred to as ranking. On the ecommerce platform, the products are ranked in multiple ways – Popularity, Price – Low to High, Price – High to Low, New Arrivals, Discount – High to Low, Fastest Shopping Time, etc.

Some of these methods of sorting or ordering are one dimensional. For example, arranging by low to high price, showing the product with lowest price on the top and maximum price at the bottom. The price is already in the data and sorted accordingly before displaying on the web pages. This feature enhances the shopping experience by allowing buyers to quickly find products that meet their price criteria. However, the complexity lies in ensuring that sorting algorithms efficiently handle large inventories and diverse pricing structures. Additionally, sorting by price alone does not account for other crucial factors like product popularity, reviews, or recent trends, all of which can impact the overall shopping experience.

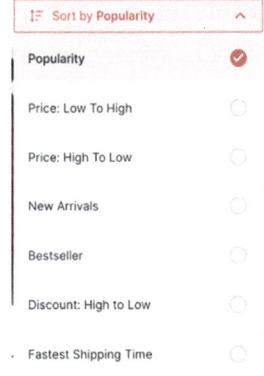

Figure 6-3. *Sorting options on an ecommerce*

Source: https://www.nykaafashion.com/

One of the options of ranking is the arranging based on popularity and is available across platforms – Amazon India, Myntra, or Nykaa Fashion, etc. Why is this option available? How do you define this popularity?

Arranging products based on popularity is complex due to the multifaceted nature of the term "popularity." Popularity can be measured by various metrics, including the number of clicks a product receives, its growth in visibility, or its overall sales volume. These metrics often do not align perfectly; for instance, a product with high click-through rates (CTR) may not translate into high sales if it does not convert well, while a product with strong sales may not have received as many clicks but could have higher buyer satisfaction and alignments. Additionally, factors such as lower-priced items often garner more views, and promotional activities can significantly impact conversion rates.

Machine learning models can balance these elements, enabling a comprehensive view of product popularity. And this is called popularity score or relevance score. This composite popularity score is then used to rank products effectively on ecommerce platforms, ensuring that the most relevant and influential products are highlighted on product listing pages.

As demonstrated through Manisha's search and navigation journeys, a vast array of products competes for top placement on ecommerce platforms. With only 40 or 50 slots available on the first page, selecting the most relevant products requires an effective algorithm. This is crucial, as buyers typically explore only a few pages before making decisions. Figure 6-4 illustrates how click-through rates (CTR) decline exponentially across subsequent pages, highlighting the importance of ensuring that the most pertinent products are prominently featured on the initial pages.

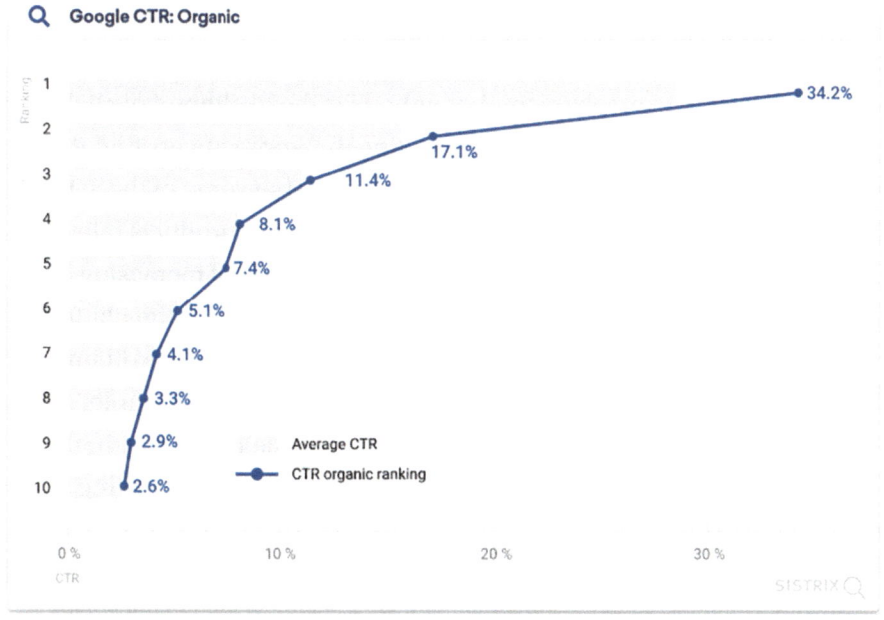

Figure 6-4. Ranks and click through rate (CTR)

In web searches, the click-through rate (CTR) experiences a notable decrease from the first position (34.2%) to 7.4% for the fifth position [7].

Similarly, both the click-through rate and the conversion rate decrease exponentially as the product position shifts, and also from the first page to subsequent pages on the ecommerce platform. This trend may be attributed to the presence of relevant products at the top, and/ or potentially causing browsing visitors to disengage if they fail to find relevant products early in their journey.

The goal is to show the products at higher positions that will increase the likelihood of clicks and conversions. However, only a select few products can attain top positions. This is pivotal for both buyers, who desire the most relevant products, and the ecommerce platform, which seeks to promote products that boost sales and conversions.

When Manisha is unregistered or not logged in, her choices and preferences are less understood. However, the platform cannot simply choose products randomly for each position. Instead, it should display products based on a mechanism that enhances the likelihood of clicks and sales. If products are popular, indicating that more buyers find them relevant and useful, there is an increased chance that some of these buyers share similar preferences with Manisha, making those products relevant to her as well. So, the popularity improves the chances of increasing relevance to buyers and also delivering better outcomes for the platform.

In the ecommerce landscape, product popularity can have different shades based on distinct performance characteristics. For example, there are products with high order counts, attributed to their longevity in the market and competitive pricing below the platform's average selling price. These items, often low-ticket, benefit from substantial discounts. Conversely, newly launched product series, endorsed by brands and promoted on social media platforms like Instagram, garner high clicks but comparatively lower conversion rates. Despite their popularity, these products yield lower order volumes. Luxury brands and products, on the other hand, boast higher-priced items with fewer sales, yet generate substantial revenue due to their premium nature.

Selecting the most suitable products for individual buyers in light of these diverse scenarios is a complex task. It requires a comprehensive consideration of multiple factors such as pricing, discounts, promotions, and brand appeal. Relying on a single factor alone is insufficient to determine popularity of the products; popularity is a multi-dimensional measure in ecommerce.

The relevance of popularity is determined by evaluating all products on an ecommerce platform. Popularity is assessed by considering factors such as performance, potential, and efficiency, which collectively contribute to determining the popularity of each product.

The performance metrics encompass factors such as unit sales, sales value, and clicks. However, when products receive greater visibility and higher placement on product listing pages, their performance metrics may be inflated due to positional bias or systematic advantages. Hence, performance measures alone may not be good enough.

A few indicators may signal negative performance, for example, a high buyer-return rate for a product. The high return rate might be associated with various factors, including fit issues, quality concerns, or delayed delivery. In assessing popularity, product return performance metrics should be considered, particularly those attributed to fit or quality issues. The only positive performance-based popularity can show the products with high return rate up in the list and that leads to increase in the operational costs.

Efficiency measures gauge how effectively a product contributes to driving performance metrics. These measures include ratios such as click-through rate, cart add rate, and conversion rate, reflecting the product's ability to deliver outcomes when given the opportunity. However, efficiency metrics can be misleading when based on small sample sizes. For instance, if a product is viewed by only five visitors and receives one order, the conversion rate appears high at 20%. Yet, this may not provide sufficient evidence to support the efficacy of the metrics. Therefore, it's essential to consider performance measures alongside efficiency metrics to ensure a comprehensive evaluation of a product's effectiveness.

New arrivals often lack established performance and efficiency metrics. However, these products play a vital role in infusing freshness onto the platform and sparking interest among potential buyers. Additionally, well-known brands possess their own captive audiences and hold aspirational appeal among visitors. Therefore, factors such as newness, brand recognition, along with clicks and views, can serve as indicators of great potential for these products.

Moreover, when a brand previously exclusive to a competitor's ecommerce platform becomes available on a platform, certain products from that brand may capture the attention of users. Despite the absence of performance metrics on the ecommerce platform, indicators such as product ratings, reviews, and rankings on the competitor platform can offer insights into their potential popularity. Therefore, these metrics can be factored into assessing the popularity of these products on the ecommerce platform.

Lastly, but certainly not least, diversity and novelty are crucial considerations when showcasing products on an ecommerce platform, especially within the realm of fashion ecommerce. The inclusion of diverse products and the introduction of novel products not only enrich the shopping experience but also cater to a broader range of tastes and preferences among buyers. This emphasis on diversity and novelty helps keep the platform dynamic and engaging, encouraging continued exploration and discovery for shoppers [8].

Ranking Function

A method for combining multiple attributes into a composite factor used for ordering products in commerce is referred to as ranking [1]. In ecommerce, ranking is a function of performance, potential, and efficiency characteristics that can be used for rearranging products. These factors are integrated into a rank score that determines how products are ordered and displayed. This can be mathematically represented as

Rank Score = f(Performance, Potential, Efficiency)

In ecommerce, ranking products involves assessing performance, potential, and efficiency to ensure relevant products are prioritized. Performance metrics such as product views, orders, and sales amount reflect current success and popularity. Potential indicators, like emerging growth in views and the newness of products, gauge future prospects and increasing interest. Efficiency measures, including click-through rates and conversion rates, evaluate how well products convert opportunities into actual sales. By integrating these factors, ecommerce platforms can create a comprehensive ranking system that balances immediate success with future potential, optimizing product visibility and relevance.

The functional form for ranking can vary from simple linear models to more complex algorithms, depending on the specific ranking system employed. Linear models use straightforward calculations to determine ranks, often based on weighted sums of performance metrics. In contrast, complex algorithms may incorporate advanced techniques, such as machine learning or neural networks, to analyze and rank products based on a multitude of factors and interactions. The choice between linear and complex models depends on the goals of the ranking system and the sophistication required to achieve accurate and relevant product placements.

In summary, the ranking function can be constructed as either deterministic or predictive. Before discussing the details of a ranking functional form, it is useful to provide a historical perspective on ranking and its applications beyond ecommerce.

Ranking – A Brief History

The application of ranking has become ubiquitous in our digital lives, influencing our choices and decisions across a wide range of online activities. Their history dates back to the 1940s, when the concept of ranking emerged in the field of economics. For decades, ranking systems have been widely used to evaluate colleges and universities [3]. However, it was the development of the PageRank algorithm by Google in 1998 [2, 4, 5] that truly propelled ranking algorithms into the mainstream.

Google PageRank – Revolutionizing Search

The PageRank algorithm, which forms the backbone of Google's search results [2], revolutionized the way we retrieve information online. By analyzing the links between web pages, PageRank assigns a ranking score to each page, determining its relevance to a user's search query.

"PageRank works by counting the number and quality of links to a page to determine a rough estimate of how important the website is. The underlying assumption is that more important websites are likely to receive more links from other websites" [9].

The algorithm's success inspired other technology giants to incorporate ranking algorithms into their platforms, leading to a diverse range of applications.

Ranking algorithms have infiltrated various aspects of our online experiences, shaping our interactions with technology platforms and influencing our decision-making processes. Here are a few notable examples, in addition to Google Search and PageRank:

- **Bing Ranking**: Microsoft's Bing search engine utilizes a sophisticated ranking algorithm that considers user relevance, query context, and document quality, ensuring users receive relevant and satisfactory search results.

- **Facebook's Feed Algorithm**: Facebook's algorithm determines the most relevant stories for users based on likes, shares, and comments, prioritizing content that is not only engaging but also resonates within the user's social circle, influencing their online interactions and social experiences.

- **Amazon Product Ranking Model**: Amazon's algorithm uses factors such as the number of reviews, average rating, and product price to ensure products are ranked in a manner that aligns with user preferences, guiding them toward products that are likely to meet their needs and expectations.

These examples illustrate the versatility and adaptability of ranking algorithms, which have become indispensable tools in shaping user experiences and influencing decision-making processes across the digital landscape.

In the realm of ecommerce, ranking algorithms play a pivotal role, orchestrating the order in which products are displayed, influencing user behavior, purchasing decisions, and ultimately, sales success. Unlike classification or regression tasks, ranking involves arranging items in an optimal and relevant order based on a variety of factors.

Product ranking plays a crucial role in various aspects of the ecommerce experience. Here are some common applications:

- **Search Results Ranking:** When a user enters a search term, products are displayed based on their ranking scores. This ensures the most relevant and appealing products appear at the top.

- **Product Recommendations Reordering:**
  Recommended products are often reordered based
  on a ranking score that considers user preferences,
  product popularity, and past interactions. This
  personalized the shopping experience and increases
  the likelihood of conversion.

- **Category and Brand PLPs (Product Listing Pages):**
  Products on category and brand pages are arranged
  based on their ranking scores. This helps users navigate
  through a large selection and find what they're looking
  for more efficiently.

Search Ranking in Ecommerces

When Manisha searches for "printed sarees" on an ecommerce platform,
the search functionality matches the search terms with the products. The
platform has around 6000 sarees meeting the search terms. The search
page displays 40 products on the first page before showing pagination
for exploring additional search results. The ranking algorithm enables
the platform to select the most relevant 40 products for the search results
landing page. The objective is to improve the relevance of these products
to the user for a given search term.

Considering the limited information about Manisha's preferences
and tastes, using popularity as an approach to show products could be
effective. The assumption is that if products are popular, they are liked by
many buyers, which increases the chances of these products aligning with
her preferences as well.

Ecommerce search enables buyers to discover the products they need
from the vast ocean of options available on the platform. Even the smallest
improvements in speed and search result quality can have a significant
impact on revenue and user experience [6].

The search process involves two stages: the retrieval of products (Search) and then displaying the products based on popularity ranking to improve user experience and business impact, such as click-through rate and conversion rate. The retrieval of relevant products for search terms, covered in the previous chapter, is the initial stage.

Figure 6-5 visually illustrates the two key stages involved in creating an effective search journey on an ecommerce platform.

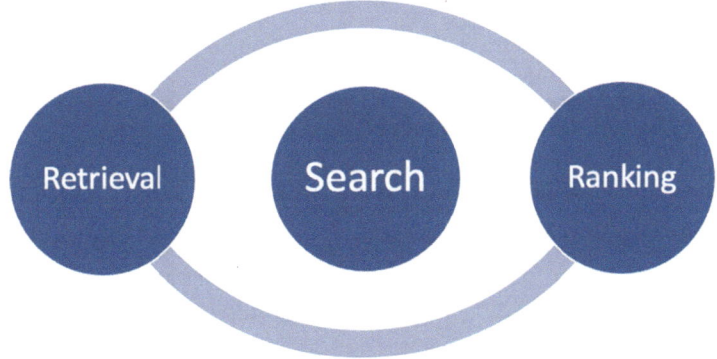

Figure 6-5. *Role of ranking and retrieval in search*

Ranking products based on their popularity is significantly important. Popularity is multi-dimensional, involving various performance, efficiency, and product characteristics for measuring product popularity.

Similar to Manisha, thousands of users are searching for "printed sarees" on the ecommerce platform. From the retrieved 6000 printed sarees, the top 40 most popular products based on historical data are displayed. These products are selected based on characteristics such as high search click-through rates, better conversion rates, higher clicks, and orders. This approach ensures that the most relevant and desirable products are shown to users, enhancing their shopping experience and increasing the likelihood of purchase.

The popularity score can be modeled using either deterministic or probabilistic approaches, incorporating performance, potential, and efficiency characteristics of products. Probabilistic models, often utilizing machine learning techniques, can capture the complex interplay of these factors and deliver more nuanced outcomes. Common machine learning methods used for popularity ranking include XGBoost rankers and deep learning-based ranking models, both of which enhance the accuracy and relevance of product rankings.

Ranking: A Deterministic Model

Performance factors like product views, orders, and sales value, along with efficiency attributes such as click-through rate, cart add rate, and conversion rate, could be considered for deriving a composite score using a ranking function. These composite scores are then used for ranking the products when users like Manisha search for printed sarees.

One of the simpler forms of the ranking function could be a deterministic approach. In this method, weights are assigned to each of the performance factors and efficiency attributes to develop a composite score. The weights are determined based on the business context and criteria, arrived at after discussions with relevant business stakeholders. This approach allows for a clear and consistent ranking of products based on predefined priorities and objectives.

Ecommerce platforms like Flipkart run the "Big Billion Days" sale every year, during which more discounted products are displayed to users. In the deterministic-based composite ranking score approach, higher weights are assigned to the percentage discounts on the products to reflect their increased attractiveness.

Similarly, during festivals, new and trending products are highly sought after. In the ranking scores, higher importance is given to new arrivals and trendy products to match user preferences and seasonal trends. This adaptive weighting ensures that the most relevant products are highlighted based on the current context and user interests.

In the deterministic approach, weights are assigned to pre-identified factors based on their importance, and the composite popularity scores for products are calculated accordingly. This method can be complex in practice, as products can appear on various pages, including brand, category, or search results, and any changes in weights affect all these listings. During sales or seasonal events, the weights of relevant factors are adjusted to highlight more pertinent products. For instance, while newness might typically account for 5% of the score, it could be increased to 10% during sales to better represent new arrivals.

A deterministic approach for ranking products is a simple and transparent method for creating composite scores to arrange products, such as in ecommerce search results. However, it comes with some challenges and limitations.

Ecommerce platforms, like Myntra or Amazon, handle hundreds of thousands of search query terms, each associated with numerous products to arrange. As discussed in the search chapter, search terms are normalized to aggregate similar intent queries into the same normalized query. For example, "T-shirt for men" and "Men's t-shirt" can be grouped into the same normalized query.

However, managing the weights assigned to each normalized query can be quite demanding. Additionally, the product performance and efficiency attributes used are based on historical data, not necessarily reflecting how they will influence future product performance. This limitation makes it difficult to adapt to changing trends and user behaviors dynamically.

Ranking: A Machine Learning Model

A supervised machine learning framework can be utilized to create ranking scores within each normalized search query. The ranking aims to improve business metrics and user experience.

Labeled ranks are essential for developing a model that ranks products. One method for assigning ranks to products for each major search query is based on manual review and business knowledge. However, given the volume of search queries, this approach can be challenging and a bottleneck.

A systematic and automated method for assigning ranks to products for each search query uses a functional form. This form can combine total sales, order count, and conversion rate. Weights can be assigned across products and search queries. While the order count may bias toward low-priced items due to their high volumes, considering the conversion rate priorities products recommended more frequently or displayed to a wider audience, ensuring that popularity is not solely determined by price.

$$Ranks = f(GMV, Orders, Conversion\ Rate)$$

Weights for these three factors are determined through simulation and analysis, taking into account the varying measurement scales of the factors. Scaled values are used before applying weights to derive the final values and ranks.

The ML-based predictive modeling structure employed for the ranking model involved creating ranks for three days and utilizing a previous 30-day window to generate features.

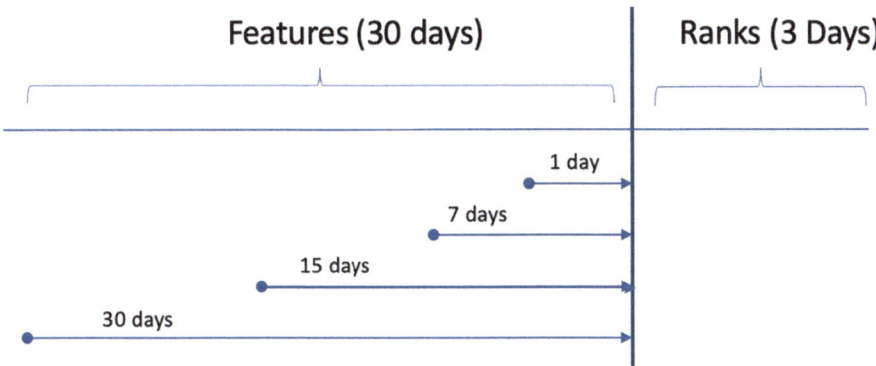

Figure 6-6. *Structure of supervised ranking model development*

To construct these features, a variety of product performance drivers and product attributes were employed. The following base features were considered:

- **Product Characteristics:** Color, Fit Type, Pattern, etc.

- **Performance Statistics:** Orders, sales value, product views, etc.

- **Efficiency Ratios:** Cart Add Rate, Conversion rate, etc.

- **Feedbacks:** Return Rate, ratings, reviews

Product review data on ecommerce platforms provide valuable insights that can enhance popularity ranking models. However, since reviews are textual, they must be converted into numerical features before being integrated into the ranking model.

Common methods for feature extraction include calculating sentiment scores, such as percentages of positive and negative sentiments, and converting these into ranked values (e.g., whether a product falls in the top 20% based on positive sentiment). Additionally, natural language processing (NLP) and transformer models can identify the top tags associated with products. Features such as the count of reviews and positive sentiment scores for these tags are also considered. For instance, if a tag frequently highlights product design issues, products associated with that tag might be ranked lower.

In deep learning-based ranking systems, textual reviews can be transformed into embeddings using advanced techniques. These embeddings capture the semantic meaning of the reviews, allowing the model to use them directly as input. This approach enables the ranking system to leverage the nuanced information contained in the reviews, such as overall sentiment and context, to improve the accuracy and relevance of product rankings.

The predictive ML model approach leverages the historical features for finding relevant ranks in the period ahead.

Case Study: Ranking Search Results for Fashion Products
Context

Indian fashion ecommerce platform wanted to develop a seamless and personalized shopping experience for its buyers. Over 30% of buyers used the platform's search function, with a majority clicking on products displayed in the results. A crucial aspect of this experience was effectively ranking search results for fashion products, ensuring buyers saw the most relevant and appealing items.

The diverse and dynamic nature of fashion products, encompassing a wide range of styles, brands, colors, sizes, and price points, made it challenging to accurately assess their relevance and appeal to individual buyers. The ranked list also needed to align with the latest styles and preferences.

Target Variable

Designing an ML-based ranking system required labeled data. However, manually ranking product lists for each search term was impractical and costly due to the platform's vast product volume.

Based on the 3-day period, product sales value (GMV), conversion rate, and click through rate for the search journeys were considered for these products. Based on the business discussions, 60,20, and 20 weights were considered across these three factors to arrive at the ranking of the products within each search query.

Features

For creating features, product performance features were created across the last 90 days separately for the search journeys and across all journeys. For these variables, time-variant features are created to capture trends and patterns.

In addition to performance and popularity features, product intrinsic features like sales reviews, ratings, and returns were also considered.

Model

A gradient-boosted tree algorithm with a list-wise ranking objective function was used for developing the ranking of the model. This approach is called Learning to Rank Model.

Evaluation

The model was evaluated using metrics like Precision/Recall @5 and 20, and mean reciprocal rank (MRR). Evaluating the model on 10% of platform traffic revealed significant improvements in business metrics, with CTR increasing by 2% and conversion rate by 0.2%.

Ranking Algorithms: Learning to Rank (LTR)

Learning to Rank (LTR) is a widely used technique in real-world systems for generating ordered lists as a subsequent stage of search and recommendations on ecommerce platforms. LTR stands out among various base rankers due to its versatility and effectiveness. It employs a global scoring function to assess the relevance of each item in a candidate set, producing an ordered list based on these scores.

There are different ways to formulate a loss function, which quantifies the difference between predicted rankings and the true rankings. The choice of an appropriate loss function is essential for training the model effectively. Three functional forms – point-wise, pairwise, and list-wise – are typically used in Learning to Rank models. Each has its own advantages and challenges.

Point-wise learning to rank (LTR) approaches focus on minimizing the individual error between each item's predicted relevance score and its actual relevance score.

Pairwise learning to rank (LTR) takes a different approach. Instead of focusing on individual items, it turns the ranking problem into a binary classification task. This involves creating pairs of items from a given ordered list (learning examples) and labeling each pair based on their rank.

Both point-wise and pairwise methods have limitations as they do not consider the entire ranked list in their loss functions, leading to suboptimal rankings.

The list-wise approach addresses these limitations by considering the entire list of ranked items as input and utilizing multivariate loss functions to determine the optimal ordered list as output. This holistic approach takes into account the relative ordering of items, resulting in more accurate and meaningful rankings.

List-wise ranked lists can be evaluated by a set of metrics. This approach measures the quality of the ordering by considering the entire ranking list. These metrics are particularly relevant in ecommerce, where users typically focus on the top few results in the list.

- Mean Average Precision (MAP) summarizes the precision at all possible cut-off points.

- Mean Reciprocal Rank (MRR) measures the average reciprocal rank of the first relevant item in the ranking list, providing a simple and efficient way to assess ranking quality.

- Normalized Discounted Cumulative Gain (NDCG) considers the position of relevant items within the ranking list, with higher positions assigned greater weight.

Finding relevant products at the top of search results or recommendations significantly impacts both the buyer experience and conversion rates. As discussed previously, ensuring that the most relevant items are prominently displayed can enhance user satisfaction and drive higher sales. NDCG (Normalized Discounted Cumulative Gain) specifically measures how well the most relevant products are positioned at the top of the list. It accounts for the importance of relative positions by applying a discount that decreases exponentially as the rank of relevant items goes lower. This approach ensures that the metric reflects the greater significance of top-ranked items in influencing buyer decisions and achieving better business outcomes such as conversion or GMV.

The power of ranking algorithms extends beyond ecommerce search. They also play a vital role in rearranging recommendation lists on ecommerce platforms, tailoring them to individual users for a more engaging shopping experience. In the next section, we delve deeper into the development of ranking models specifically for recommendation systems, exploring techniques and strategies to optimize product recommendations and drive user engagement.

Ranking for Recommendations on Ecommerce

For enhancing the user experience and driving sales, ecommerce platforms employ recommendation engines that suggest relevant items to each user. These recommendation engines typically consist of two phases: candidate generation and ranking of the candidates.

The candidate generation phase involves identifying a subset of items from the vast catalog that are potentially relevant to the user. This process often utilizes recommendation engine algorithms, with deep learning frameworks gaining popularity due to their ability to handle diverse data types and scale efficiently.

The ranking phase involves prioritizing the candidate items based on their relevance to the user. This is where ranking algorithms come into play. Ranking algorithms analyze user data, item characteristics, and contextual factors to determine the order in which items should be presented to the user.

Figure 6-7 illustrates the steps involved in presenting video recommendations on YouTube.

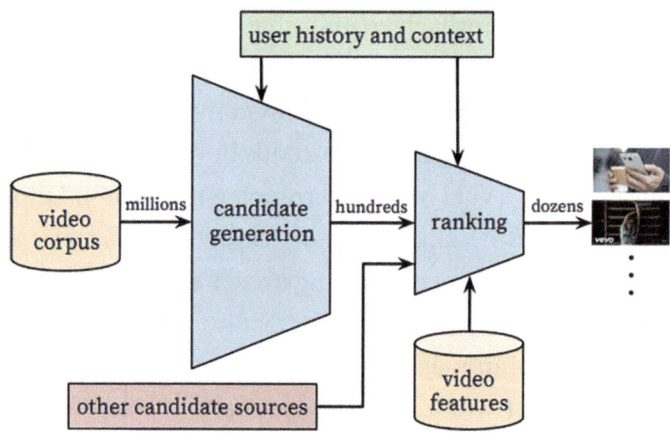

Figure 6-7. *Recommendations and ranking*

Source: https://dl.acm.org/doi/pdf/10.1145/2959100.2959190

Ranking for Similar Product Recommendations

In the world of online fashion, "similar product" recommendations aren't just window dressing – they're a conversion-driving powerhouse. Buyers browsing product pages (PDPs) rely on them like a trusted stylist, to discover their preferred products and make purchase decisions. While only 10–20% of Indian fashion ecommerce visitors actively engage with these recommendations on the detailed product pages, they're high-intent buyers, primed to buy. That's why optimizing the order of these suggestions is paramount for boosting conversion rates.

User expectations for "Similar Products" are driven by both the product's inherent features like style, material, and color, and its visual elements like patterns, shapes, and colors. Similar Products recommendation engine does product feature matching and extracts visual cues from images, and factors in product popularity data to dynamically generate relevant product recommendations. This ensures users discover items that not only match in product styles but also align with current trends, ultimately leading to a more satisfying and engaging shopping experience.

Similar product recommendations can be refreshed regularly and effectively, even when product characteristics (such as color or pattern) remain static. This is accomplished by updating implicit factors derived from buyer interactions and purchase patterns, which are captured in both the user feature matrix and the product feature matrix. As buyer behavior evolves, these updates ensure that recommendations adapt dynamically to changing preferences over time. Furthermore, machine learning-based ranking systems integrate updated product metrics to continuously refine the products shown to buyers. This approach ensures that recommendations remain relevant and aligned with current trends and performance data, even in non-personalized contexts.

On the ecommerce platform, buyers see five "Similar Products" recommendations below their chosen products with an option to explore more similar products, as shown in the figure below. However, most engagement happens with the default list displayed. Therefore, the primary goal is to ensure the top five products in this widget are the most relevant and impactful, maximizing business outcomes. This involves designing ranking algorithms that optimize the placements of these top products for each buyer.

Figure 6-8 illustrates a layout example for similar product recommendations.

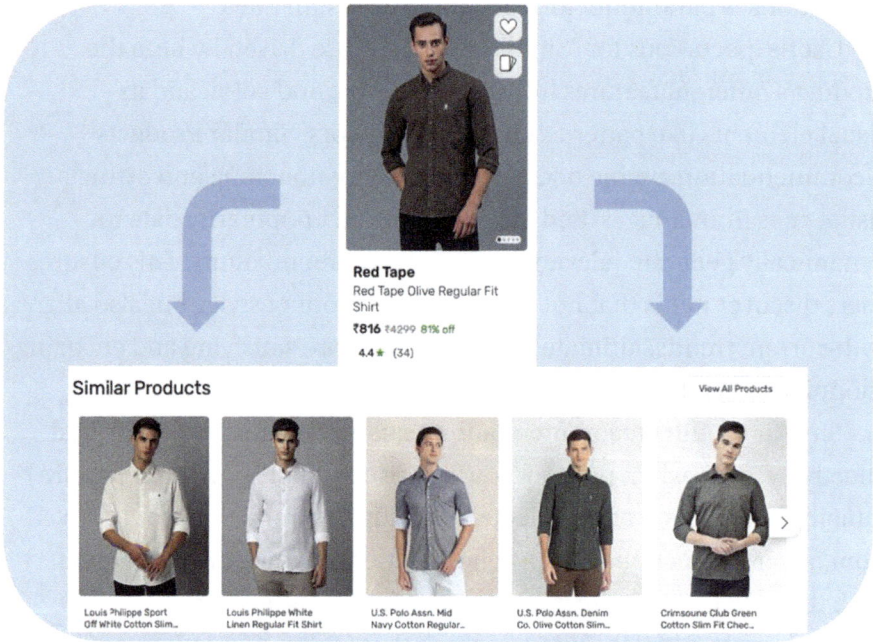

Figure 6-8. *Similar product display layout*

Candidate lists of similar product recommendations are generated using image-based Similar Product Generation – identify items with similar visual elements, such as patterns, colors, and shapes, etc. and

Product Attribute-based Similar Product Generation – incorporate features based on product descriptions and specifications like material, size, or brand.

The candidate similar products are rearranged using Learning to Rank (LTR). The model prioritizes products with high recent sales, user ratings, and click-through rates, ensuring buyers see products that are not only relevant but also well-received by others.

Learning-to-Rank (LTR) models in ecommerce face several challenges due to the dynamic nature of the industry. In fashion ecommerce, for example, trends change rapidly, and new designs can quickly become popular. However, these new products may lack historical data, making it difficult for LTR models to effectively rank them. Data quality is crucial; biases or inaccuracies in the data can significantly impact the model's outcomes. Furthermore, deep learning-based ranking models, while powerful, can suffer from overfitting, especially if they are not adequately regularized or if they rely too heavily on historical patterns.

For example, when a user browses a pair of chunky sneakers on the ecommerce platform, the user is shown options that are trending and aligned with the current preferences, leading to a more satisfying shopping experience.

Conclusion

Ranking is fundamental for enabling buyers to find relevant products and for users to discover information on digital platforms. Whether it's an ecommerce platform, search engine, or streaming service, the strategic placement of products, content, or services through ranking significantly enhances the user experience.

Google's pioneering PageRank algorithm underscored the potency of ranking systems, revolutionizing how we interact with information. Today, ranking is an indispensable component of digital success, demanding strategic approaches and continuous optimization.

Ecommerce thrives on relevance. By presenting the most pertinent products at the forefront, platforms enhance the buyer experience, drive conversions, and foster loyalty. The importance of ranking is evident across touchpoints, from showing a ranked list of options on the homepage to displaying ranked search results and browsing pages.

As the digital landscape evolves and ecommerce platforms take center stage for millions of buyers, ranking algorithms play a crucial role in sorting products or options on these platforms. Personalized category sequences and widget options on the homepage engage users and improve their experience. During the search journey, ranked product lists improve click-through rates (CTR) and conversions. Search ranking is enabled through machine learning (ML) and artificial intelligence (AI) algorithms. Similarly, recommendations and products on listing pages are augmented using ranking algorithms on top of identified candidates.

In the next chapter, the focus is on how personalization elevates the buyer experience. In addition to personalized ranking across search and recommendations, personalization influences buyer journeys on ecommerce platforms through layout personalization and personalized offers and promotions, improving buyer experience and business outcomes.

References

[1]   Wikipedia (2023), Ranking, Retrieved from https://en.wikipedia.org/wiki/Ranking

[2]   Wikipedia (2023), PageRank, Retrieved from https://en.wikipedia.org/wiki/PageRank

[3]   Wikipedia (2023), College and university rankings, Retrieved from https://en.wikipedia.org/wiki/College_and_university_rankings

[4]   Amy N. Langville, Carl D. Meyer, Google's
        PageRank and Beyond: The Science of Search
        Engine Rankings, PRINCETON UNIVERSITY
        PRESS, https://gi.cebitec.uni-bielefeld.
        de/_media/teaching/2019winter/alggr/langville_
        meyer_2006.pdf

[5]   Massimo Franceschet, PageRank: Standing on the
        shoulders of giants, Aug-2010, https://arxiv.org/
        pdf/1002.2858.pdf

[6]   Constructor Team, The 8 Best Papers on eCommerce
        Search Algorithms, Mar-2020, https://constructor.
        io/blog/the-8-best-papers-on-ecommerce-search-
        algorithms/

[7]   Dave Chaffey, 2024 comparison of Google organic
        clickthrough rates (SEO CTR) by ranking position,
        Jan-2024, https://www.smartinsights.com/
        search-engine-optimisation-seo/seo-analytics/
        comparison-of-google-clickthrough-rates-by-
        position/

[8]   NEIL HURLEY, MI ZHANG Novelty and Diversity
        in Top-N Recommendation – Analysis and
        Evaluation, Mar-2011, https://dl.acm.org/doi/
        abs/10.1145/1944339.1944341

[9]   Paul Covington, Jay Adams, Emre Sargin, Deep
        Neural Networks for YouTube Recommendations,
        Sep-2026, https://dl.acm.org/doi/
        pdf/10.1145/2959100.2959190

Personalization – AI-Crafted Customer Experience

Introduction

Products on search and category navigation pages are ranked to enhance their relevance to buyers, leading to increased engagement and improved business outcomes such as higher conversion rates and gross merchandising value (GMV). While ranking is generally influenced by product attributes and emerging trends, performance can be significantly enhanced through personalization, where products are ranked based on individual buyer affinities. This chapter explores the concept of personalized ranking and various other use cases of product personalization.

A recommendation engine is used to achieve personalization, but not all recommendations are personalized. For example, similar product recommendations may not be personalized to start with, and an additional personalized re-ranking layer is created to personalized recommendations for each of the buyers.

R. Prajapat, *AI-Powered Ecommerce*, https://doi.org/10.1007/979-8-8688-0923-1_7

Personalization extends beyond search and recommendations, impacting every step of buyer touchpoints in ecommerce and on the ecommerce platform. In this chapter, we will explore a few scenarios and how personalization is driving performance for ecommerce.

Mohsin, who owns an apparel fashion outlet in Bangalore, has a regular buyer named Jasmin who lives in an apartment complex close by. Jasmin visits the shop and requests Mohsin to show her a few dresses for an upcoming wedding. Mohsin, being familiar with Jasmin's style and preferences, as well as exact size, selects a set of options and Jasmin picks up one that she likes the most. Jasmin navigates through a vast portfolio of options seamlessly, saving her time and facilitating faster purchase decisions. This is an example of personalized and contextual knowledge creating a smooth user or customer experience.

While Mohsin's personalized approach works well for a small number of users. He registered with a fashion ecommerce platform. The ecommerce platform has millions of daily visitors. The platform should be able to provide a personalized experience similar to Mohsin's. One approach could be that there are thousands of people similar to Mohsin who look at all the things users are doing and have done to develop the views of the users. As soon as the user logs in to the platform they start pushing options and products relevant to each of the users.

The above approach of creating manual but personalized feed for the users is not scalable and highly expensive. And the platform may not even be able to deliver a similar level of personalized experience. How artificial intelligence and machine learning can help Mohsin/ecommerce to deliver personalized experiences even when there are thousands and millions of users?

Millions of users visit an ecommerce platform daily. How can a fashion ecommerce platform accommodate the personalized needs of its millions of daily visitors?

Jeff Bezos in 1998 mentioned that if there are 4.5 million customers on Amazon platform, there should be 4.5 million stores [1] . That is the vision he had at that point and AI/ML is enabler to his vision.

Personalization in ecommerce refers to tailoring the shopping experience based on user context and preferences. In this chapter, we explore a wide array of personalization strategies and examples in an ecommerce environment and examine how they are implemented using Artificial Intelligence (AI) and Machine Learning (ML).

Personalization in ecommerce revolves around using updated buyer information to tailor product recommendations and search results. Initially, capturing all relevant details about buyers is crucial. Once collected, this information must be accessible to algorithms for further processing. These algorithms then use the updated data to refine recommendations, search results, and other features based on specific objectives.

The process of capturing, processing, and integrating data into algorithms involves various technological components and a well-designed architecture to enable effective personalization.

Consider a scenario: Alex is a frequent shopper on an ecommerce platform and searches for "blue t-shirts." His profile, which includes preferences for brands and price ranges, is used to tailor the search results. The buyer's profile is stored in a Customer Data Platform (CDP), such as Elasticsearch, and is leveraged alongside the search query results. The system performs real-time personalized re-ranking based on Alex's profile and the current search results.

Location-Based Personalization

Manisha sees a TV advertisement and visits an ecommerce platform. This is her first visit. The platform recognizes that she is from Bangalore. The home page of the website prioritizes the categories that resonate well with the Bangalore visitors; this is done based on location. Considering this is her first visit, it shows the offers and discounts relevant for the first time visitors. The offer is very aggressive and can be availed across categories and brands.

She clicks on the first option and starts her exploration on the platform. The location-based personalization and contextual (first visitor) offer entices Manisha to progress to the next steps of the journey.

She clicks on the women's Indian wear category (although it was not personalized as the system could not recognize the gender) and further explores kurta sets within the women's apparel category. She lands on the "Kurta Set" category and finds a list of kurta sets. The kurta sets are prioritized based on popularity in the region for her to review and explore the trending products. She wanders across products and adds a few to the Wishlist. She stops her exploration journey for the day, and leaves a wealth of information about her interest and preferences.

The platform captures information on what is browsed, purchased, and wish-listed across different brands and categories. This rich data becomes the key to understanding unique preferences and crafting hyper-personalized experiences [2].

For first-time visitors like Manisha, the platform personalized based on the device type (e.g., iOS vs. Android) and/or location-based personalizations. The location may not be available if someone visits a website or location tracking is not enabled.

Home Page Personalization

After a few days, Manisha returns to the ecommerce platform and lands on the homepage of the mobile application. This is her repeat visit. The ecommerce platform recognizes her based on the device cookies. The cookies can store various data points beyond browsing history, including location and device type and even though she is not logged in, the home is personalized based on the extracted data from the cookies.

A home page on the ecommerce platform has a few important elements – category navigation, hero banner (big section with attractive visuals showing casing offer), and a few other widgets or sections. But for repeat customers, the personalized widgets such as recent views, previous purchases, and recommendations for you will start piping up for the buyers.

Platform recognized that Manisha is interested in Indian wear, her preferences for Westside brands (based on the clicks in the previous visits), and her price affinity is above average. The system maps her preferences and pulls the hero banner images tagged with these characteristics. A banner visual that highlights website brands along with a visual of a lady wearing a kurta-set. Also, the sequence of the category in the category carousels is personalized for her. The home page contents are adopted based on her interest, making it exciting for her.

In addition to personalized hero banners and categories, personalized widgets are visible on the home page, including "Recently Viewed Products" and "Products Inspired by Browse History." These widgets are created with the aim of establishing continuity from the previous visit(s) to help her with the purchase journey. Machine learning algorithms identify an additional list of products that are typically explored or purchased when users browse the products that Manisha had browsed in the previous visit(s). These products help her explore additional but relevant and personalized choices available on the platform.

The two personalized widgets are the examples of product recommendations, but personalization on ecommerce goes much beyond product recommendations. The landing page layout, contextual search, and showing curated offers and promotions are few of the additional dimensions of personalizations on the home page for ecommerce. In the dynamic layouts, the colors and layout of the home page is adjusted based on the buyer preferences, creating a familiar and inviting space that feels like it was made just for them. In the contextual search, search terms are understood considering previous browse and purchase behaviors. For example, Intelligent search powered by machine learning understands and maps "kurta" search term to women kurta-set based on browsing history, showing more of woman kurta to Manisha based on her clicks for the category instead of man kurtas.

For Manisha, the full home page is personalized, showing the recently viewed products, additional options that are aligned with previous journeys and the offers are customized for herself. This creates a smooth purchase journey for Manisha and engages her toward product discovery.

A homepage, or digital front door, is crucial in ecommerce for creating an engaging and personalized experience. When buyers visit the homepage, it can be tailored specifically to them, with each widget personalized based on their profile and preferences.

For instance, if a buyer is a male from Southern India, the homepage can be customized to feature product categories and recommendations relevant to this demographic. This process involves mapping the buyer's demographic and geographic details to a category performance matrix. This matrix, akin to the user-item interaction matrix, links category performance with buyer segments. It is decomposed into a category feature matrix and a buyer segment feature matrix.

To personalize the homepage, each buyer is assigned to a segment based on their profile. The feature affinities of the buyer segment are then compared to the category feature matrix. This comparison identifies similar categories, which are then ranked to ensure the most relevant categories appear on the homepage.

Additionally, every widget and feature on the homepage – such as banners, product suggestions, and promotional content – can be personalized to enhance the buyer's experience. By dynamically adjusting these elements based on the buyer's profile and interactions, the ecommerce platform can improve engagement and satisfaction.

Now, we will explore a case study of how a leading fashion ecommerce platform in India, Myntra, has personalized its storefront. How have they created a personalized experience for their users and what is the impact of the revised journey?

Case Study: Tailoring StoreFront at Myntra *[3, 23]*

A leading Indian fashion ecommerce platform, Myntra is redefining the shopping experience through hyper-personalization. A sophisticated AI engine that transforms their homepage (called "storefront") into a curated fashion for each individual customer.

Myntra goes beyond basic data gathering to understand the essence of its customers' "fashion taste." Customers' "fashion taste" encompasses not just preferred brands and categories but also their subtle inclinations toward specific product attributes like colors, fabrics, and even preferred fits. This rich tapestry of preferences is considered in the personalized storefront.

Myntra's AI engine deciphers personal style through purchased products, extracting key features like preferred brands, colors, fabrics, and styles. Features linked to each banner are arrived at based on clicks on the banners and target products. By matching customer "fashion taste" with the essence of each banner, Myntra's AI orchestrates a storefront symphony where every ad speaks directly to individual taste.

"With this change, two users who enter the Myntra store on their apps at the same time see a completely different store, tailored to meet their fashion shopping needs" [3].

The homepage of an ecommerce platform features various widgets, such as banners, recommendation sections, and call-to-action (CTA) journey widgets, each containing specific content. The goal is to dynamically arrange these widgets to tailor the homepage experience for individual users. To achieve this, a reinforcement learning algorithm known as the Contextual Multi-Armed Bandit is employed to generate personalized layouts for each buyer.

The platform utilizes a federated architecture to minimize latency, ensuring fast and responsive interactions. The ranking services are managed via an in-memory cache, which helps to further reduce response times. Additionally, the system integrates both near real-time and batch data pipelines to balance the freshness of data with the need for comprehensive processing. According to performance metrics, the p99 response time for this service is 50 milliseconds, indicating a highly efficient and responsive system.

Personalization is an experience to engage customers across the touchpoints and deliver outcomes for ecommerce business. Personalization improves conversion for the user visits and chances of them coming back. In the next section, we explore the impact of personalization in detail.

Impact of Personalization in Ecommerce

Personalization has woven itself into the very fabric of ecommerce. Its roots go back to the dawn of online shopping, with pioneers like Amazon and eBay leveraging their power in the 1990s. Fast forward to the early 2000s, and recommendation engines became go-to tools for personalization. Amazon's collaborative filtering algorithms, subtly nudging customers toward the next purchase based on their past behavior.

Enter the 2010s, when the landscape shifted dramatically. Artificial intelligence (AI) and machine learning (ML) became the new sheriffs in town, enabling more accurate predictions and real-time adjustments. Personalization moved beyond product recommendations, and was used for personalized emails, chatbots, virtual assistants, and even dynamic website content, all tailoring the experience to individual preferences.

The last few years have witnessed a meteoric rise in hyper-personalization. Every click and action is dissected and analyzed, shaping product suggestions and content in real time. Think of it as a bespoke shopping experience, curated just for a user. Technologies like natural language processing (NLP) and computer vision further refined these strategies, extending personalization to promotions, discounts, and loyalty programs.

Personalization is no longer a differentiation but essential for survival [4]. Customer expectations on personalization is on the rise – 73% of shoppers expect brands to understand their unique needs and expectations[5]. A successful personalized program can deliver significant business impact, "20 percent higher customer-satisfaction rates, a 10 to 15 percent boost in sales-conversion rates" [4]. According to McKinsey research, the companies that excel at personalization generate 40% more revenue than average players [5].

Great examples of successful personalization programs for the ecommerce platforms and across marketing funnels are described below.

- **BigBasket:** Indian Grocery Platform achieved improved click-through rates from 1.8% to 2.5%–3% and a surge (26% uplift) in addition to cart using right time push notifications to the customers [4].

- **Multi-Category Ecommerce Marketplace:** A multi-category ecommerce platform selling products fashion, footwear, accessories, electronics, and luxury verticals leveraged personalization and achieve 2X revenue growth and 35% increase in customer engagement [7].

- **Tata CLiQ:** Tata CLiQ is an omni-channel marketplace owned by Tata Unistore Limited, a part of the Indian conglomerate Tata Group. Leveraged personalized message for cart abandonment campaign, leading to 1.5X boost in Click Through Rate (CTR) [8].

- **Paytm:** Indian Financial services company, Paytm, personalized recommendations on its home page and observed a 5.5–6% conversion rate from the Paytm Mall homepage [9].

185

- **Sephora:** Sephora, founded in 1970 and acquired by LVMH in 1997, has ~2300 stores in 33 countries worldwide. Sephora is investing in AI for personalized product recommendations [10].

These are just a few published studies that highlight the importance of personalization. Personalization is at play across touchpoints in ecommerce.

With the advent of deep learning and large language models, the nuanced details of buyer behavior – such as feedback, clicks, and purchases on ecommerce platforms – are captured and represented as embeddings. Vector databases have further enhanced the ability to leverage these details almost in real time, enabling highly personalized search results and other buyer-specific features on ecommerce platforms. The challenge of creating and managing a large number of features for machine learning models has been mitigated by the power of deep learning, which can effectively handle hundreds of features and scale efficiently.

In this era of hyper-personalization, the constraints of cluster or segment numbers are no longer a limiting factor. Personalization efforts are increasingly focused on individual buyer levels, allowing for a more tailored and precise buyer experience.

In the example of Manisha, she visits again organically, but the ecommerce platform cannot rely solely on visitors remembering to return. The marketing team, responsible for driving customer engagement from existing customers or visitors, typically known as the Relationship Management (CRM) team, works on designing campaigns to bring back visitors to the platform. Personalized email marketing can significantly improve key performance metrics, including open rates (the percentage of recipients who open the emails), click-through rates (the percentage of recipients who click links within the email to visit the e-commerce platform), and transaction rates (the percentage of recipients who ultimately make a purchase on the platform) [6].

Personalization in Marketing

Manisha was engrossed in an ecommerce platform when she received a call from her office, abruptly halting her purchase journey. Similar to Manisha, 95–98% of the customers drop off from the purchase journey and do not purchase any products for the visit on ecommerce platforms.

Only 1–3% of visitors end up purchasing products, resulting in a conversion rate of around 1–3% for fashion ecommerce. The platform sends a notification to Manisha a few hours later, reminding her about the products she viewed but did not purchase. Manisha clicks on the notification, which takes her back to the products she viewed. This is a success of notification to remind and create interest for Manisha to re-engage with the platform.

The marketing team focuses on setting up notifications to enhance customer re-engagement efforts. These campaigns run with various messages and promotional offer constructs to improve the response for the notification campaigns. Whether visitors will respond or click on notifications and reach the platform is not a one-size-fits-all scenario but a multi-dimensional one. A machine learning-based framework helps the marketing team identify what works for each user. This helps in improving notification response rates or click-through rates.

Manisha resumes her product discovery journey, explores various kurta-sets, mostly printed and embroidered and adds products to the cart! This is a significant event on the purchase journey, cart add. Around 5–10% ecommerce visitors reach this stage. This is called cart add rate or add to cart rate, percentage of visitors adding at least one product to the cart. She leaves the product/s in the cart and ends her purchase journey.

A significant count of users, like Manisha, add products to the cart but do not purchase anything during the visit. Research suggests that the shopping cart abandonment rates in India are around 51%, and some of the reasons are unexpected shipping cost, pay on delivery option, product price changes, high product price as compared to other ecommerce platforms, lack of product reviews, etc [14, 15]. Again, the marketing team leverages customer relationship management system and process to reach to these users for re-engaging them may be via notifications.

Not all notifications are guaranteed to reach customers, which is why the CRM team leverages multiple channels to reach out to visitors. Some of these channels include SMS, WhatsApp, etc. While notifications are a low-cost channel, SMS and WhatsApp incur some costs (typically 12–15 P per communication). Despite the low cost per communication, the overall cost can escalate significantly considering millions of users on the ecommerce platforms. Who are the users that should be targeted and what channel and how many times?

The marketing and data science teams collaborate to identify the preferred communication channels and determine the offer or product communication to buyers. Buyer segmentation and machine learning algorithms help the marketing team improve campaign response rates by identifying who clicks and returns to the platform, ultimately leading to more purchases. Common intervention touchpoints include product views without adding to cart or added products to the cart but not purchased. These customers are reached out to via multiple channels such as email, SMS, or WhatsApp.

Each channel in a marketing strategy has distinct response rates and associated costs. Achieving business outcomes, such as increasing visits through CRM campaigns while managing costs, requires careful consideration. An optimization framework can assist the marketing team by determining the optimal volume of contacts to make across each channel, aiming to maximize responses while adhering to budget constraints. This approach ensures effective resource allocation and cost management while striving to achieve the desired marketing goals.

Case Study: User re-engagement powered by Machine Learning (ML) Model

Context: For an ecommerce platform, over 2 million daily visitors. A significant percentage of these customers explore products. However, only 8–10% of these customers add products to the cart, and many of them do not return by the next few days. The platform invests heavily in marketing efforts to attract users to the platform, leading these visitors on their own is not a right choice. The focus is on developing marketing interventions to re-engage users who explored products but did not add them to the cart or make a purchase.

Objective: Developing a machine learning model to predict the likelihood of a customer returning when a product is promoted or included in an SMS/WhatsApp campaign.

Approach: For model development, users who have browsed at least two products in their latest session are selected. The historical data is used to label products as 1 if they were explored and purchased, and 0 otherwise. User browsing behavior, such as clicks on brands or categories, and time spent on each product, are used as features. A machine learning model is then developed to identify products that have the highest likelihood of bringing back users to the platform.

Outcome: Improved email response rate for the campaign that has personalized products for each campaign.

An SMS reminds Manisha about the Kurta-set she intended to purchase. She opens the ecommerce app and instead of opting for a browsing journey, she prefers the search navigation journey this time.

Search Personalization

Manisha leverages the search bar on the ecommerce application and types "kurta-set." The search suggests the search term as "kurta set for women," and she starts her product purchase journey.

When the search engine returns the products, the list is personalized considering that she has already explored kurta sets with printed and embroidered patterns. This makes shopping experience seamless for Manisha as the ecommerce search factors in her preference toward a certain type of patterns. The personalized search experience for Manisha not only considers the pattern but also various subtle preferences such as price range, brand, occasion, or fabric.

Search personalization leads to faster product discovery for the customers and improved shopping experience, leading to increased conversion rate.

Up to 30% of customers on ecommerce platforms harness the power of search to find what they're looking for, generating over 50% of sales [13]. Optimizing and personalizing search journeys directly impacts both business success and customer experience. Research has shown that the position of a product in the search results significantly impacts its Click-Through Rate (CTR) [11]. This is where personalized ranking comes into play. By considering individual customer data like past purchases, browsing behavior, and contextual factors (location, time of day, etc), AI-powered ranking models can curate product listings that resonate deeply with each shopper.

Airbnb, a pioneer in personalization, employs short-term signals (like recent clicks and searches) and long-term memory (previous bookings and preferences) to craft deeply personalized search results using cutting-edge embeddings. This approach has resulted in a remarkable 20% increase in CTR [12].

In addition to search results, home page, and marketing personalizations, there are numerous features on ecommerce platforms that can benefit from personalization. We all appreciate when results are curated and crafted for our needs, tastes, and preferences.

Before we delve into other examples of personalization on ecommerce platforms, let's explore a whole new world of personalization: the story of Stitch Fix. Stitch Fix is a company that designs and delivers fashion products to customers even before they place an order.

Design to Delivery Personalization: Stitch Fix – A Personalized Stylist

Stitch Fix is a leading online clothing subscription service-based company in the United States, founded in 2011. It takes personalized experience in ecommerce to the next level by bringing algorithms and humans together. With customer information around price, size, style, lifestyle, and other preferences, the company designs personalized styles for each buyer and delivers products along with a personal note [16].

In a nutshell, imagine a carefully curated box of five fashion items, handpicked just for a customer by a virtual stylist who knows every detail relevant to the buyer, just like a personalized stylist. This isn't magic; it's the power of data science and human stylists combined [17].

Stitch Fix leverages a treasure trove of data about customer style preferences, fit, and past purchases to curate a selection that feels effortlessly personal. It's like having a fashion fairy godmother, whisking away the guesswork and leaving you with a box of delights to try on at home. Whether you keep all five treasures or simply send some back, the experience is about discovering new favorites and refining personal style. The free returns create a closed loop to further refine the algorithms and understand more about customers' tastes and preferences.

Stitch Fix's business model is fundamentally centered around personalization, creating a unique and tailored experience for each customer. On the other hand, most ecommerce platforms augment their features with personalization to deliver a better customer experience and achieve positive business outcomes.

Manisha spends significant time exploring interesting trends on Instagram and engaging with the personal network. She notices an interesting product advertisement (Paid Social campaign) and clicks on the banner. This redirects her to the ecommerce product detailed page (PDP). She explores the products and clicks on the similar product widget.

Personalized Similar Product Recommendations

Users like Manisha can reach the product detail page (PDP) through various ways, such as searching for a product and clicking on a product from the search results, navigating via brand and product listing pages, browsing through category hierarchies, or directly from a product link shared in SMS, social media, or social media campaigns.

Users who land on the product pages are a high-intent visitor cohort, with a significantly higher conversion rate for this group. To improve engagement rates for these customers, the platform provides a number of widgets on the product detail page. One of these widgets is "similar products," which helps users explore relevant products and make purchase decisions.

Similar products construct identifies candidate products that are like the product explored by Manisha. In the recommendation chapter, we illustrated the approaches that help in generating a list of products that share similar characteristics. These algorithms create a non-personalized set of products for all the users, irrespective of the current user preferences [18].

Based on the previous browsing and purchases history for Manisha, it is evident that she prefers printed and embroidered patterns, above average price products and light color products. The non-personalized product list from the similar product recommendations must be re-prioritized for Manisha based on her preferences.

Case Study: Personalized Similar Product Recommendations at Myntra [18]

Context: *Myntra is a leading fashion ecommerce platform in India. Product discovery is important for the platform to improve user engagement. Similar product recommendations not only make it easier for users to discover relevant products but also contribute to increasing the time users spend on the platform, thus improving overall engagement. Personalized recommendations can further improve user experience and engagement.*

Objective: Re-arrange product according to each user, considering user's preferences and fashion tastes for the similar product recommendation widget.

Approach:

- *Utilizing implicit data such as clicks, views, and purchases, low-dimensional latent space vectors are generated for both users and products through matrix factorization methods.*

- *Candidate generation involves creating non-personalized recommendations of similar products by comparing the current product with the product catalog.*

- *User vectors encapsulate user interests and preferences within low-level embedding vectors, with analogous features captured in product vectors within the same space.*

- *Products (candidates) are then re-ranked based on both product–product similarity scores and product–user similarity scores.*

Evaluation and outcome: *The count of products displayed to users remains unchanged but are reordered according to user affinities. Performance evaluation utilizes metrics such as Precision, Recall, and Mean Average Precision based on the top 15 products. From a practical perspective, A/B testing is commonly used to assess the impact of improvements in machine learning algorithms on real buyers. It is essential to measure both technical metrics, such as recall and precision, but significantly more important to measure business-facing metrics, such as click-through rate, average order value, and conversion rate. Evaluating these metrics provides a comprehensive view of how changes affect buyer experience and overall business performance.*

The personalized similar product recommendations are driving user engagement on the product pages (PDPs) and improving the add-to-cart rate. The add-to-cart rate typically ranges from 5–10% and varies by category, location, and other factors.

Users add items to the cart or wish-list, and sometimes these products go out of stock. When a product goes out of stock for a user, the platform can leverage the "personalized similar product" functionality to identify alternatives that could resonate well with the user and reach out to the customers using SMS or WhatsApp channels.

From the cart page, users proceed to the next steps of the purchase journey such as selecting delivery address and making payments. A wide range of payment options ensures that users can leverage the options most relevant to them. The lack of relevant payment options could be one of the factors for cart abandonments [14, 19].

A long list of payment options is available in India and across the world. List of payment methods available on Amazon India and Tata CLiQ are as follows.

Figure 7-1 provides a glimpse of the payment page layout on two ecommerce platforms.

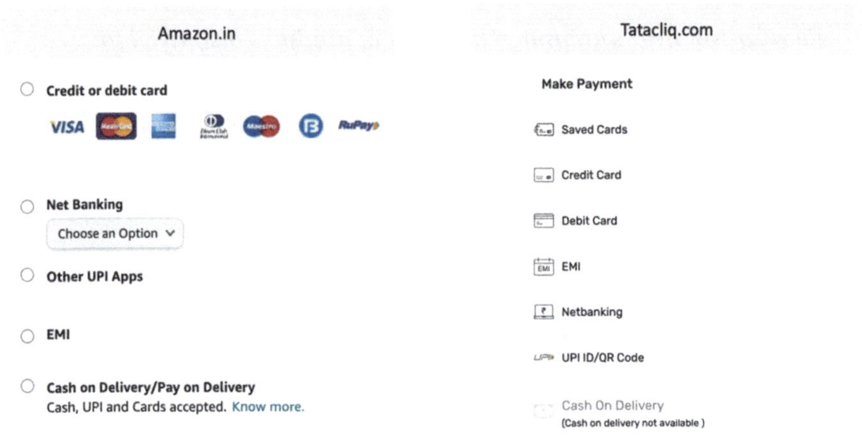

Figure 7-1. *Payment options on e-commerce*

In India, there is a persistent reliance on cash on delivery (COD) for ecommerce payments. However, credit/debit cards have gained traction in urban hubs and among tech-savvy demographics, contributing around 29% to ecommerce payments. Nevertheless, card and net banking penetration remains low in smaller cities and towns in India [20].

Additionally, the Unified Payments Interface (UPI) is witnessing rapid adoption, further diversifying the payment ecosystem in India [21] Affinities toward payment options vary significantly across cities and towns in India.

The sequence of these payment options has an influence on the customer experience and conversion rate. Studies reveal that there is around a 30% increase in conversion when offering users their top 3 preferred methods [22]. The clear payment option affinities across towns and cities underscore the importance of adopting various payment options across different regions. Showing user level payment options can further improve the customer experience.

Machine learning empowers this personalization by dynamically predicting preferred payment methods based on a wealth of data. Past purchases, geographic location, user demographics, browsing behavior, and even engagement with discounts and promotions can all inform and influence the right payment sequences.

Machine learning-enabled payment mode personalization prioritizes Cash on Delivery payment for first-time users, especially from smaller cities or towns. For repeat users, it considers previously used modes of payment and even selects the most relevant option as the default payment mode.

The personalized payment sequence for each user goes beyond merely providing payment choices. By leveraging data-driven insights and tailoring the checkout experience to individual user preferences, platforms can significantly reduce cart abandonment, drive conversions, and foster a more seamless and satisfying ecommerce journey. The case study provides an overview on how payment sequence personalization influences customer behaviors and a positive impact on the business outcome.

Case Study: Payment Sequence Personalization for Repeat Users

Context: *For an Indian fashion ecommerce platform, the cash on delivery (COD) usage was around 40% and it incurs cash handling fees. Also, usage of top 3 payment methods stood around 20–25%. Focus was to improve customer experience by showing relevant payment options first and also increase adoption of pre-paid payment methods such as Cards, Net Banking, and UPI.*

Objective: *Developing personalized payment sequences to show most relevant payment options on the top.*

Approach: *For repeat users, their past payment method usage is taken into account as input for the machine learning-based prediction of their next payment mode. The recency and frequency of each payment mode are utilized for creating features in the model.*

Outcome: *For users who were shown personalized payment options, there was an increase in prepaid payments (+3%), and now three times as many customers were using one of the top three payment options shown, leading to an improved customer experience.*

Customizing payment options for individual users' preferences not only boost conversion rate but also improves buyer convenience.

Preferences for each payment mode can be influenced by buyer's historical behavior (considered in the above scenario) and promotional offers. For example, a buyer might typically use UPI for payments but switch to a credit card if an additional discount is offered. This requires real-time personalization because the terms and conditions of offers can change frequently. If a buyer has already used a credit card offer, the system should adjust future rankings to reflect this, potentially lowering the prominence of the credit card option. With multiple dimensions to consider – such as offer eligibility, transaction history, and current promotions – real-time payment mode personalization becomes complex but essential for optimizing the user experience and maximizing the effectiveness of promotions.

From displaying region or geo-based trending personalized products to individual usage-based payment sequences, personalization is critical for ecommerce platforms. In this chapter, several different personalization use cases are discussed. When a user lands on the home page, a personalization welcomes and sets the stage for an engaging journey with relevant brands, promotions, and products, along with a personalized layout. As users navigate through their shopping journey, the platform adapts to their choices. From personalized recommendations to payment options, each of these aspects caters to the tune of user preferences.

On the journey, when users leave the platform with unfinished shopping, the platform re-engages them with personalized marketing campaigns to entice them to restart their purchase journey. Personalized experiences also lead to higher repeat purchases from users. These personalizations drive user engagement, conversions, and have a significant impact on user experience and business performance.

Once, the personalized experience enables purchases, the focus shifts to meeting customer experience on delivering the right product and on time. The post-order journey is crucial for both users and ecommerce businesses, and optimizing delivery operations can make a significant impact on business success.

Summary

Personalized marketing campaigns attract more buyers to ecommerce platforms cost-efficiently. Once buyers are on the platform, personalization enhances engagement and conversions.

A personalization framework enables ecommerce platforms to deliver personalized content, search results, product recommendations, and promotional offers. Personalized content, such as homepage banners and promotions tailored to buyer location, browsing history, and preferences, significantly improves buyer engagement.

Buyer engagement is further influenced by personalized search results and improved relevance ranking of products on listing pages based on browsing patterns and other behavioral data points. Deep learning-based personalization frameworks deliver scalable personalization on ecommerce platforms.

Personalized payment options, offers, and promotions enhance conversion rates. Additionally, personalized messaging and communications help reduce cart abandonment. Machine learning (ML) methods are instrumental in achieving these personalizations.

The next chapter will delve into enhancing buyer experience after order confirmation on an ecommerce platform and increasing delivered revenue (Net Merchandising Value – NMV). Reducing the gap between Gross Merchandising Value (GMV) and NMV is crucial, as it reflects the efficiency of the processes involved. Cost management and operational efficiency are critical for business success. This chapter will explore the key drivers and components of efficiency in ecommerce, with a particular focus on improving delivery rates and minimizing buyer returns.

References

[1]   Leslie Walker, Amazon Gets Personal With E-Commerce, Nov-1998, https://www.washingtonpost.com/wp-srv/washtech/daily/nov98/amazon110898.htm

[2]   Seekmeai, How AI Impacted E-commerce: Personalization, Recommendations, and Customer Experience, Oct-2023, https://medium.com/@seekmeai/how-ai-impacted-e-commerce-personalization-recommendations-and-customer-experience-cdb7ec64faf8

[3]   Anand Agrawal, Personalizing the shopping experience on Myntra Storefront, May-2028, https://medium.com/@_anandagrawal/personalizing-the-shopping-experience-on-myntra-5c7eb1f04138

[4]   Shana Haynie, Ecommerce Personalization: What It Is, Examples, & How to Build an Effective Strategy, Feb-2024, https://www.moengage.com/blog/ecommerce-personalization/

[5]   Mckinsey, Unlocking the value of personalization at scale for operators, Feb-2022, https://www.mckinsey.com/industries/technology-media-and-telecommunications/our-insights/unlocking-the-value-of-personalization-at-scale-for-operators

[6]   Ted Vrountas, 68 Personalization Statistics Every Digital Advertiser Must Keep in Mind, https://instapage.com/blog/personalization-statistics

[7]   vue.ai, Multi-category marketplace improves conversion with 1:1 personalization, https://vue.ai/resources/case-studies/ecommerce-personalization-platform-casestudy-for-marketplace-categories/

[8]   Clevertap, Case Study – Tata CLiQ India's fastest-growing e-commerce brand achieves 159% boost in revenue via multi-channel engagement, `https://clevertap.com/case-study/how-tata-cliq-relies-on-personalization-and-real-time-communication-for-a-4x-boost-in-ctr/`

[9]   AWS and PayTM, Paytm Boosts Homepage Sales with Personalized Recommendations Using Amazon Personalize, 2022, `https://aws.amazon.com/solutions/case-studies/paytm-personalize-case-study/`

[10]   LG Student, Sephora and Artificial Intelligence: What does the future of beauty look like?, Nov-2018, `https://d3.harvard.edu/platform-rctom/submission/sephora-and-artificial-intelligence-what-does-the-future-of-beauty-look-like/`

[11]   Hema Yoganarasimhan, Search Personalization Using Machine Learning, Jul-2020, `https://faculty.washington.edu/hemay/search_personalization.pdf`

[12]   Mihajlo Grbovic, Haibin Cheng, Real-time Personalization using Embeddings for Search Ranking at Airbnb, Jul-2018, `https://dl.acm.org/doi/10.1145/3219819.3219885`

[13]   Abigail Bosze, 23 eCommerce Site Search Statistics for 2024, `https://www.doofinder.com/en/blog/ecommerce-site-search-statistics`

[14]   Mrs. Manjula N Mr. Mahesh Kumar V, E-Commerce
Cart Abandonment: Exploring Consumer Behavior's
and Reasons for Cart Abandonment, Sep-2019,
https://www.ijert.org/research/e-commerce-
cart-abandonment-exploring-consumer-
behaviors-and-reasons-for-cart-abandonment-
IJERTV8IS090052.pdf

[15]   Laura S. Egeln, Julie A. Joseph, Shopping Cart
Abandonment in Online Shopping, Feb-2012, https://
digitalcommons.kennesaw.edu/cgi/viewcontent.cgi
?article=1006&context=amj

[16]   Stitch Fix, https://www.stitchfix.com/
personal-stylists

[17]   Blake Morgan, 5 Lessons In Personalization
From Stitch Fix, Feb-2021,https://www.forbes.
com/sites/blakemorgan/2019/12/10/5-
lessons-in-personalization-from-stitch-
fix/?sh=3fd03f24b4e8

[18]   Pankaj Agarwal, Sreekanth Vempati, Sumit Borar,
Personalizing Similar Product Recommendations in
Fashion E-commerce, Jun-2028, https://arxiv.org/
pdf/1806.11371

[19]   Axtrics, How to Improve the Conversion Rate by
Offering More Payment Options, https://ecommerce.
axtrics.com/how-to-improve-the-conversion-rate-
by-offering-more-payment-options/

[20]   Devaraj M, How Does Cash on Delivery (COD) Dominate Indian E-Commerce, Oct-2022, `https://linkedin.com/pulse/how-does-cash-delivery-cod-dominate-indian-e-commerce-devaraj-m/`

[21]   Mark Stiltner, FIVE LOCAL PAYMENT METHODS MERCHANTS NEED FOR ECOMMERCE IN INDIA, Jul-2022, `https://www.rapyd.net/blog/five-local-payment-methods-merchants-need-for-ecommerce-in-india/`

[22]   ACI ENTERPRISE PAYMENTS PLATFORM, What is the link between payment methods and conversion rates?, `https://www.aciworldwide.com/blog/what-is-the-link-between-payment-methods-and-conversion-rates`

[23]   Sanjeev Kumar Dangi, Myntra Store Federated Architecture Part 1: Automatic Layout Widget Reordering, Mar-2024, `https://medium.com/myntra-engineering/myntra-store-federated-architecture-part-1-automatic-layout-widget-reordering-f5beb4989f27`

CHAPTER 8

Efficiency a Key Enabler for Delivery Experience and Profitability

Introduction

Kavya, an eagle-eyed analyst in the category operations team of a thriving ecommerce company, spends her days meticulously reviewing transactions. One particular day, a curious case piqued her interest. A customer had placed multiple orders exceeding 50 products in the last 24 hours. Intrigued, Kavya delved deeper.

Her investigation, aided by the analytics team, unearthed a startling truth. Over 80% of these high-volume orders resulted in cancellations, failed deliveries, or incorrect returns. This anomaly hinted at a potential web of inefficiency plaguing the system.

© Ramgopal Prajapat 2024
R. Prajapat, *AI-Powered Ecommerce*, https://doi.org/10.1007/979-8-8688-0923-1_8

Discussions with colleagues unraveled a series of scenarios threatening the smooth flow of ecommerce. Some customers used nonsensical or incomplete addresses, leaving delivery executives helpless and customers uncontactable. Cash on Delivery (COD) transactions were particularly vulnerable to this tactic.

Another peculiarity emerged. Certain pin codes in Rajasthan and Uttar Pradesh displayed a significantly higher rate of order cancellations, primarily associated with UPI payments. In a different case, a seller flagged a concerning rise in "wrong returns." Upon investigating return patterns across pin codes and delivery partners, a startling connection surfaced. These fraudulent returns originated from a specific pin code near Surat, Gujarat, and were linked to a few delivery executives. It appeared to be a case of collusion between buyers and delivery partners, exploiting loopholes in the system.

These are just a few examples from the vast landscape of Indian ecommerce, highlighting the constant battle against inefficiencies and platform abuses. Buyers, delivery partners, and even some seller team members can exploit vulnerabilities within the platform and its processes.

Canceled order, a failed delivery, wrong delivery, and a fraudulent return add cost to ecommerce operations and disrupt operations, frustrate customers, and erode seller trust. Imagine the ripple effect: a disgruntled buyer leaves negative feedback, impacting seller reputation and potentially deterring future purchases. Additionally, wasted resources like packaging materials, damaged products, and logistics inflate operational costs.

Ecommerce platforms deploying AI enabled Kavya and her team to bring forth some of these platform abuses and in a number of scenarios, Kia (an intelligence bot), Kavya's AI colleague, stops platform abusers placing orders on the platform and also validating return requests of the buyers on the platform.

Order confirmation through to delivery or product retention involves multiple stages and processes, with various stakeholders, including buyers, sellers, and logistics companies, playing crucial roles. Managing this complex ecosystem within a marketplace model is vital for maintaining a competitive edge and operational efficiency. Data intelligence, predictive alerts, and preventive machine learning models are essential for optimizing these processes.

In this chapter, we will explore the order fulfillment journey in detail, focusing on the role of AI and ML in enhancing efficiency, buyer experience, and profitability. Key challenges in the ecommerce order delivery journey include

- Buyer Cancellations

- Seller Cancellations

- Non-Delivery of Orders/Return to Origin (RTO) Orders

- Fake Returns or Damaged Returns

- Product Returns

We will discuss each of these challenges and explore how machine learning and AI frameworks can address them effectively.

The Ecommerce Maze: Navigating the Order Fulfillment Journey

The moment a buyer makes a payment – "confirm order" – it marks the beginning of an exciting journey for the customer's purchase. But behind the scenes, complex actions and steps unfold within the ecommerce platform. This post-confirmation odyssey, known as order fulfillment, determines how swiftly and smoothly a product reaches its final destination – the customer's doorstep – or embarks on a return trip.

Post order confirmation, the order is assigned to a customer. The seller verifies inventory, picks and packs your order with care, and hands it to a courier or delivery partner. The chosen courier partner picks up and delivers the product at the customer's doorstep. This seamless flow represents the ideal order fulfillment journey – a testament to efficient operations.

However, the path from confirmation to delivery isn't always smooth. Several roadblocks can arise, causing inefficiencies and frustration for both customers and sellers.

Figure 8-1 illustrates the various stages of the post-order placement journey for an ecommerce platform.

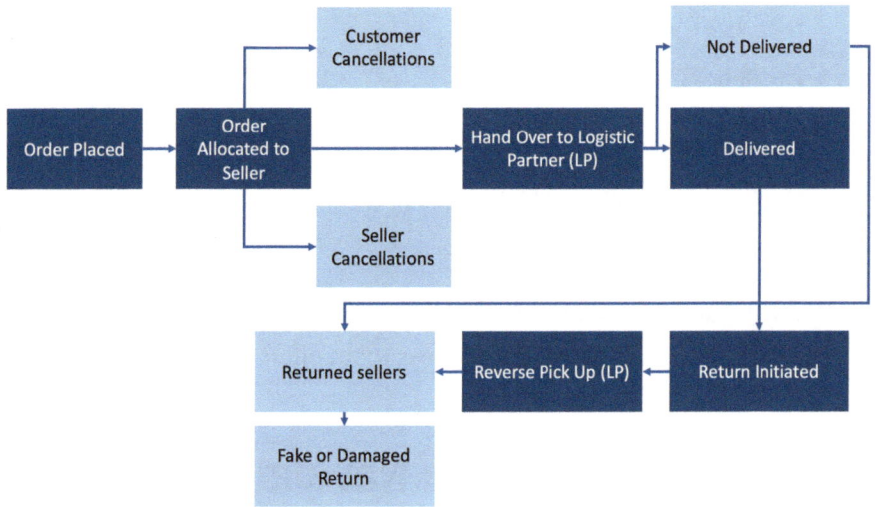

Figure 8-1. *Journey from order to delivery*

- **Customer Cancellations**: A customer might cancel an order due to second thoughts, finding a better deal elsewhere, or other reasons. Insights behind customer cancellations help in optimizing product listings and pricing strategies. Customer cancellations can stem from various issues such as buyer abuse, price discrepancies, or unexpected charges. Enhancing clarity in the buying journey can help mitigate these cancellations. Additionally, implementing machine learning-based alerts can assist in identifying and reducing instances of platform abuse related to cancellations.

- **Seller Cancellations**: Seller cancellations are often related to inventory discrepancies or pricing anomalies. Inventory issues can be mitigated through improved platform integrations that ensure real-time inventory updates. Additionally, machine learning systems can help minimize unintentional price changes that contribute to seller cancellations.

- **Not Delivered – Return to Origin (RTO)**: Return to Origin (RTO) orders are often associated with incorrect delivery addresses and buyers who are unavailable to receive their orders. In some cases, fraudulent buyers may provide false addresses with no intention of accepting the delivery. Machine learning models can be employed to identify and mitigate these issues, implementing preventive measures to reduce the incidence of RTO.

- **Customer Initiated Returns**: Change of mind, size or fit issues, wrong product, or product quality are some of the reasons the customer returns the products. Each reason for returns can be addressed through specific strategies. Fit issues can be mitigated by recommending the appropriate size based on buyer details or by implementing virtual try-on features. Quality concerns can be managed by enhancing the review process or engaging in discussions with relevant sellers to ensure product standards are met.

- **Fake Returns**: Some customers may engage in fraudulent activities by returning counterfeit or non-original products. Machine learning models can help identify and reduce these fake returns by analyzing patterns and anomalies. Additionally, implementing quality checks during reverse pickup processes can further help in detecting and managing suspicious returns.

All these can cause huge costs to ecommerce business, leading to inefficiency in order delivery journeys. This explains the gap between Gross Merchandise Value (GMV) and Net Merchandise Value (NMV). Approximately 30% of Gross Merchandise Value (GMV) is lost due to returns and cancellations [17].

Efficiency Equation

GMV-NMV = Customer Initiated Returns (CIR) + Non-deliverables (RTO)+Cancellations (Customers and Sellers)+Others

Customer Initiated Returns (CIR) refer to the value of products returned by buyers for various reasons. Ecommerce platforms generally permit returns within a specified period known as the "return window," which can vary by platform and product category. In some cases, returns may not be allowed for certain categories or products. The returns happen only after the product delivery.

The return rate, or CIR rate, is expressed as a percentage of the total GMV (Gross Merchandising Value). For instance, a CIR rate of 20% indicates that 20% of the total value of ordered products has been returned by buyers. This metric is crucial for understanding return trends and managing inventory and customer satisfaction.

Return to Origin (RTO), also known as Non-Delivery, refers to products that could not be delivered to the buyer and are instead returned to the sellers. During the delivery process, a delivery executive attempts to reach the buyer at the provided address, which includes making phone calls, if necessary. If the delivery cannot be completed due to issues such as an incorrect address or the buyer being unavailable, the product is returned to the seller (the origin).

An RTO percentage of 5% indicates that 5% of the GMV (Gross Merchandising Value) of ordered products could not be delivered and were sent back to the seller. RTO incurs costs for both forward and return shipping and negatively impacts the seller's experience and inventory management.

Customer Cancellations occur between the placement of an order and its delivery. The impact of these cancellations varies depending on the payment mode and the stage of the order journey. When a buyer decides to cancel an order, several factors come into play:

- **Prepaid Orders:** If the order has been prepaid, it may incur payment gateway charges for the ecommerce platform.

- **Order Processing:** If the seller has already started processing the order, this incurs effort and time that is effectively wasted.

- **Post-Handover:** If the cancellation occurs after the order has been handed over to the delivery partner, it involves additional delivery costs.

A cancellation rate of 5% indicates that cancellations account for 5% of the GMV (Gross Merchandising Value) of ordered products. This metric reflects the proportion of total orders that were canceled by customers.

Seller Cancellations occur when a seller cancels orders assigned to them for delivery. This can happen for several reasons, including inventory shortages or pricing discrepancies. While seller cancellations are less frequent compared to customer-initiated cancellations, they can still pose challenges in a marketplace setting. Marketplace e-commerce platforms such as Flipkart impose order cancellation fees on sellers as a measure to discourage cancellations [18].

In addition to the previously mentioned challenges, order fulfillment also incurs costs and inefficiencies due to issues such as fake returns, damaged products, and items lost in transit. Managing these issues effectively is crucial to minimizing waste and optimizing the delivery process.

Typically, 20–40% of sales values are wasted due to any of these reasons.

Each of the inefficiency drivers need to be managed effectively to improve the efficiency position.

Figure 8-2 illustrates the relationship between Gross Merchandise Value (GMV) and Net Merchandise Value (NMV), highlighting the various factors contributing to inefficiencies in ecommerce.

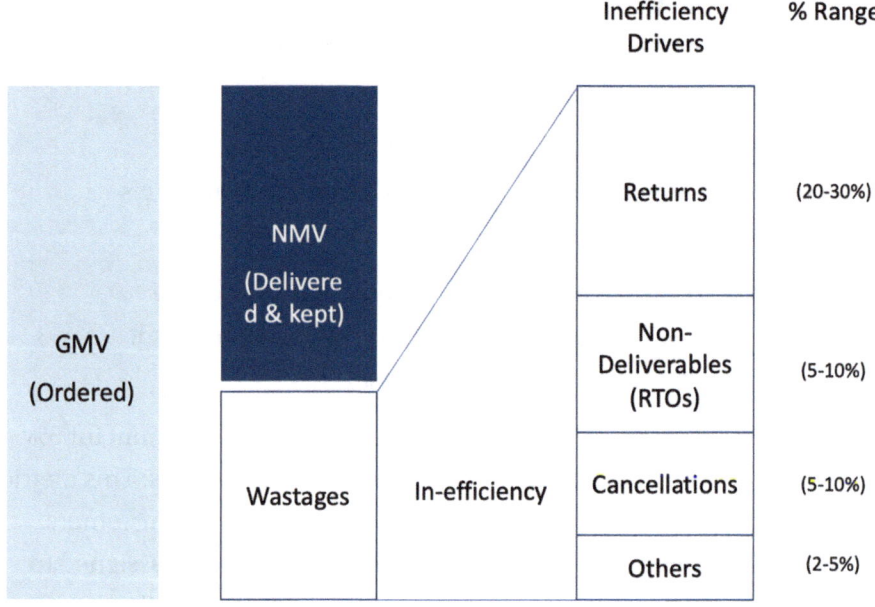

Figure 8-2. *Factors explaining gap between GMV and NMV*

Returns Orders

Ecommerce business is growing at a rapid pace and expected to surge by a CAGR of 31% from 2023 to 2029 [1] and easy return policy and cash on delivery option played significant roles. Return rate (% of orders returned by the customers) is around 25–40% in India, significantly higher compared to global average [2]. The ecommerce player may end up paying 10–15% of the order value as costs due to returns, which can significantly dent their profits [3, 4].

Order returns could be classified into two groups: returned back due to non-delivery (Return to Origin – RTO) and second, customer-initiated returns post successful delivery.

Figure 8-3 illustrates the two types of returns in ecommerce: Returns to Origin (RTO), which occur before delivery, and Customer-Initiated Returns, which happen after the delivery has been completed.

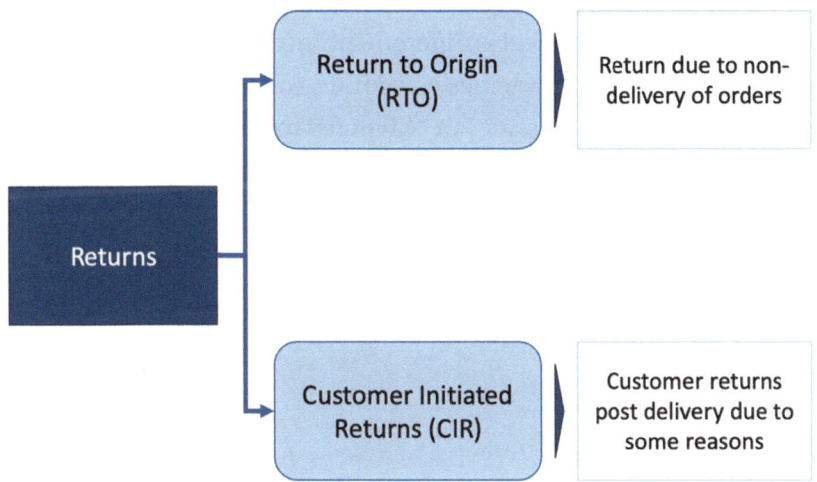

Figure 8-3. *Returns before delivery (RTO) and post delivery (CIR)*

Customer-Initiated Returns (CIR)

Once a product is delivered to the customer, the customer can return if the product is eligible for the return and it is within the return window. If the order date is Mar 13 and the return window is for seven days, the customer can initiate the return by Mar 20 only. For return, the customer needs to visit a mobile app or website and initiate the return.

Typically, the customers are asked to enter a reason for return and provide more details. Additionally, in some cases, the platform makes it mandatory for the customers to provide a product image if the return reason is a damaged product, etc.

Returns involve logistics cost (forward and reverse), packaging cost, damaged product/fake product return costs, in addition to bad customer experience.

The common return reasons for Footwear and Apparel could be related to fit issues (not feeling well), does not look good on me, quality is not satisfactory, look and feel is different, or incorrect/defective product received [4]. For home or watches categories, some of the reasons such as "part of order /product missing" or "Changed my mind" could be more common. Mix of reasons can vary by portfolio or category mix and by platform.

Figure 8-4 illustrates the key reasons for product returns.

Figure 8-4. *Product return reasons*

Return rate for apparel category could be as high as 25% of the order for a given period and return rate for watches category could be around 10–12%. For apparel and footwear categories, Size or Fit Issue is a significant contributor toward returns [12, 20].

Ecommerce platforms need to improve the processes and fortify the platform to reduce the return rate. Multiple machine learning and AI models can significantly reduce returns. For instance, the platform can offer size recommendations using historical data of products kept and returned. These recommendations can range from collaborative filtering models to advanced deep learning approaches. Additionally, showing the likelihood of returns for each product can help buyers make more informed choices. To combat abusive behavior, implementing prompts during the ordering process can also help in reducing returns.

Strategy to manage returns in ecommerce is

- **Product Onboarding:** Improving quality of product onboarded such as quality of product descriptions, image, and additional details can reduce chances of product returns.

- **Size Recommendations:** Fit issue is the single biggest contributor of product returns, especially for footwear and apparel category for ecommerce. Guiding customers with the right size decision can play a critical role.

- **Fortify Quality Control and Packaging Process:** For the marketplace model, product delivery processes involve multiple handshakes – seller operations, courier fleet, last mile delivery, and customers. Putting quality control and capturing data at various stages could be of great help in containing product returns related to damaged products or wrong product delivery.

Product Content: Creation and Validation

85% of shoppers consider product information and pictures for their purchase decisions [5]. AI tools can enable in creating product description, generating tags and designing product images [6].

For an ecommerce marketplace, there are hundreds of sellers and probably they leverage vendor teams to create, catalog, and upload products on the ecommerce platform. Considering a diverse set of resources, a multi-layered validation process is required to ensure quality content goes online.

- **Product Image Quality Validation:** Review the uploaded product images for quality and flag if any of these images are of subquality. This process can be automated using computer vision models.

- **Product Category Validation:** Leverage AI models for creating product taxonomy based on the product images and validate the product hierarchy uploaded.

- **Product Description Generator:** Based on the key product taxonomy and search keywords, Generative AI models can help in generating accurate and appealing product description.

AI-Based Size Recommendations: Enabling Customers with Right Decisions

Fit or Size recommendation is one of the key challenges in fashion ecommerce. When customers are making a purchase decision on the ecommerce platform, they need to decide on the size decision. And incorrect decision leads to customers returning products, leading to bad customer experience and increased operational cost [8, 11]. There are numerous challenges in recommending right size to a customer and some of these challenges are

- **Data Sparsity and Cold Start**: Customers can be new to platform (cold start), new to category (no earlier purchase in footwear, may have purchased apparel), and limited customers with multiple transactions across brands and categories. For example, only 0.64 co-purchase for brand–brand level, that means most of the brands are either not co-purchased or are co-purchased rarely [11].

- **Inconsistent Sizes across Brands**: Same size could refer to different measurements across brands. For example, a men's US size 8 shoe is 10 inches for a Nike trainer while an Adidas trainer measures 10.2 inches [9].

- **Multiple Purchase Personas**: For the same account (registered customer), purchases could be for multiple customer personas. Multiple personas can be identified using Gaussian Mixture Model (GMM) [9].

- **Information missing**: For footwear, the size decision is a multi-factor decision, for example, attributes like arch type, shoe width, and shoe type play a role in whether a shoe will fit well or not. As per report, most vendors on the platform only provide length of the footwear and 80% of the vendors do not provide any information on shoe width, shoe arch type etc [10].

Size Recommendations: Machine Learning (ML) Model

Size Recommendation can be modeled as Size Prediction. For example, within the Men category, the Footwear Size {UK7, UK8, UK9, UK10, UK11, UK12} can be considered as labels.

Latest transaction for a month could be considered as data for the label and all the previous history of the customers could be considered for the features. Figure 8-5 illustrates the structure used for creating features and labels in a machine learning-based predictive model.

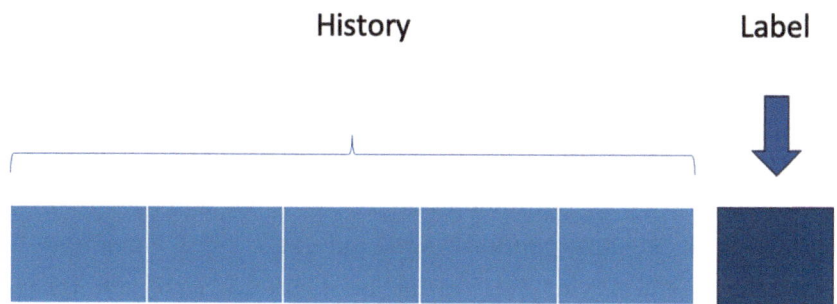

Figure 8-5. *Supervised size prediction*

Some of the features that can be considered are

- Frequent size within same category

- Most recent size within same category

- Most frequent size across categories

- Size returned within same category

- # of purchases

- Size based on similar users

- Count of purchase for each sizes (for each category and across categories)

Once data is prepared by replicating the process across multiple months (called cohorts), the data is split into training and testing samples. ML techniques such as XGBoost or Random Forest can be considered to develop ML-based size recommendations.

The fit recommendation is modeled as size prediction task as an ordinal regression problem, where the customer and product true sizes are learned by taking their differences and feeding them into a linear model [9, 13].

Size Recommendation – Skip-gram-Based Model [10]

Myntra is the largest Fashion Ecommerce company in India, it leverages skip-gram-based size recommendation. For footwear, it proposes to encode SKUs at category, gender, brand, and size levels. For example, SKU with gender "Men" article type "Casual Shoes" brand "Vans" and UK size "8" as "Men_Casual_Shoes_Vans_8." For each of these, encoding will be considered as word and multiple purchases combined as document. With these words and documents as inputs, we train the Word2Vec network. The Word2vec network would learn the links between different size labels of different brands which essentially are of the same shoe measurement

For training the model, user level purchase sequences are created by arranging purchases with the latest purchase at the end. Each of the word/ purchase is represented as a latent feature after training Word2Vec Model.

On the website, the platform asks users to provide explicit shoe preference and then find latent representation of the preference. For example, if the user has filled the shoe preference questionnaire as article type generally worn: Men Casual Shoes, preferred brand: Roadster, typical size worn: UK8, we would encode this information as a word/purchase preference and fetch the latent vector representation of this word/ preference. Based on cosine similarity, for the required category and brand combination, the size will be recommended.

At Amazon, they leverage latent features for predicting personalized size [11]. The latent factors for customers and products are related to the physical true sizes (e.g., for dimensions like length and width), and are learnt from past product purchase and returns data. Specifically, for each (customer, product) purchase transaction, our model outputs a score that is a (linear) function of the difference between customer and product true sizes, which is then used to predict the outcome. We present a construction that reduces a loss function with ordinal outcomes to a new loss function with binary outcomes. We define Hinge loss and Logistic loss variants, and propose efficient algorithms with linear time complexity for computing customer and product true size values that minimize the two loss variants.

Figure 8-6 illustrates an example of how size recommendations are presented to buyers during the size selection process.

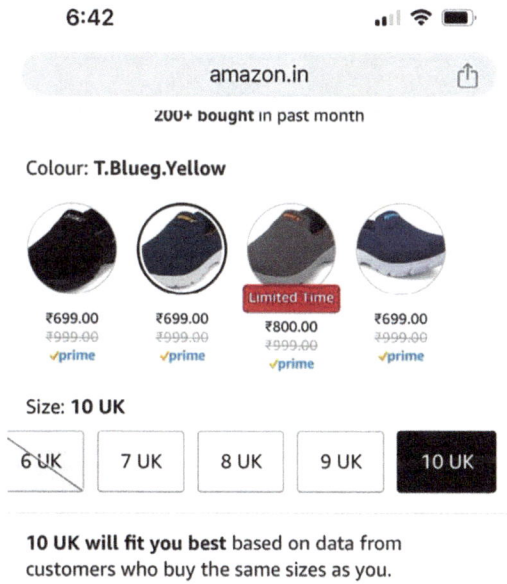

Figure 8-6. *Size recommendation example*

It also displays the message "**X UK will fit you best** based on the data from customers who buy the same size as you."

[7] Deep Learning System for Predicting Size and Fit is based on content-collaborative methodology for personalized size and fit prediction. In this approach, customer and article embeddings are created in parallel, discovering implicit properties of individual customers and articles for personalized recommendations.

Figure 8-7 illustrates the deep learning architecture employed for fit size recommendations.

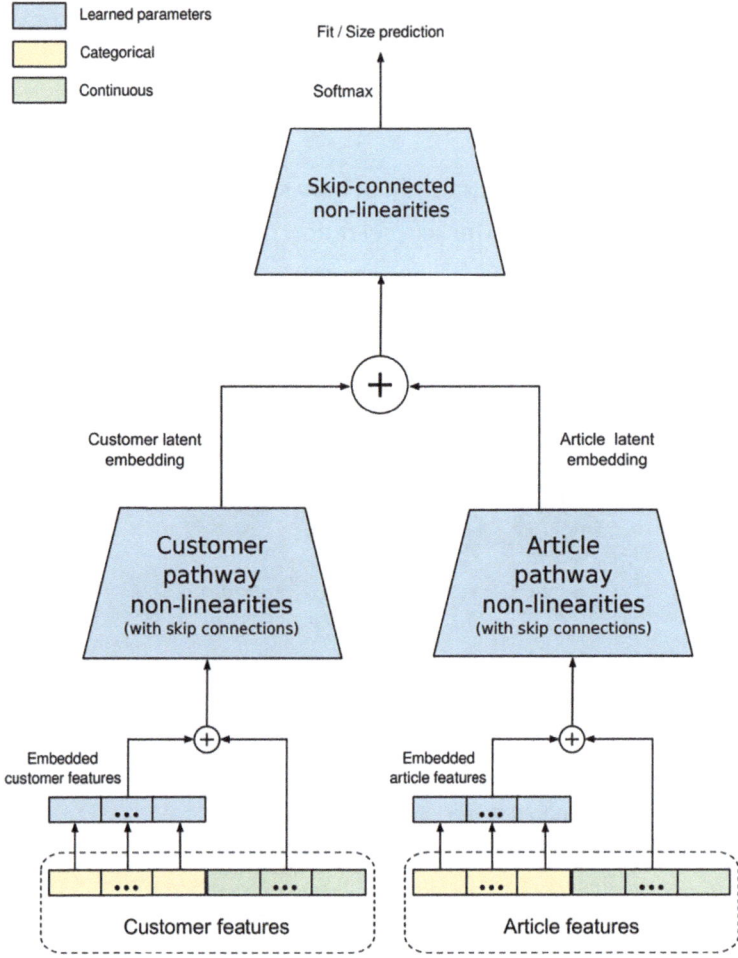

Figure 8-7. *Deep learning based size prediction*

Source: A Deep Learning System for Predicting Size and Fit in Fashion ECommerce

Standard approaches to collaborative filtering solely rely on interaction data to model customer behavior [14], but for a vast majority of customers, such data is sparse. This results in an extremely sparse customer–article interaction matrix, which makes it difficult to model preferences of every individual customer on a personalized level. Additional information in the form of customer and article attributes can, however, help to deal with the sparsity and cold-start recommendations

Recommending Clothes Sizes with Product Size Embeddings (PSE)[9]

Recommending the right size for a clothing item during online shopping remains a challenge. While traditional recommender systems suggest similar products based on past purchases, accurately recommending size requires considering a customer's body type and fit preferences. This case study explores the Product Size Embedding (PSE) model, a neural collaborative filtering approach that leverages implicit user behavior data to recommend clothing sizes.

Product Size Embeddings (PSE) model addresses this challenge by learning size-specific embeddings for each product. This approach builds upon the concept of neural collaborative filtering, where user and item interactions are used to create latent representations.

- **Implicit Signals:** PSE utilizes implicit user behavior data such as purchases and returns, eliminating the need for explicit user size information.

- **Learning Size Embeddings:** The model learns a vector representation (embedding) for each product-size combination. These embeddings capture the inherent relationship between similar body shapes and clothing sizes.

- **Asymmetric Framework:** For efficiency, the model adopts an asymmetric framework. Instead of learning individual user embeddings, it focuses on product representations, significantly reducing the number of model parameters.

- **Multi-class Classification:** PSE formulates size recommendation as a multi-class classification problem. Given a user and a product, the model predicts the most suitable size for the customer in that specific product.

Figure 8-8 visually presents how purchase history, including both purchased and non-purchased sizes, is utilized to predict the correct size for buyers.

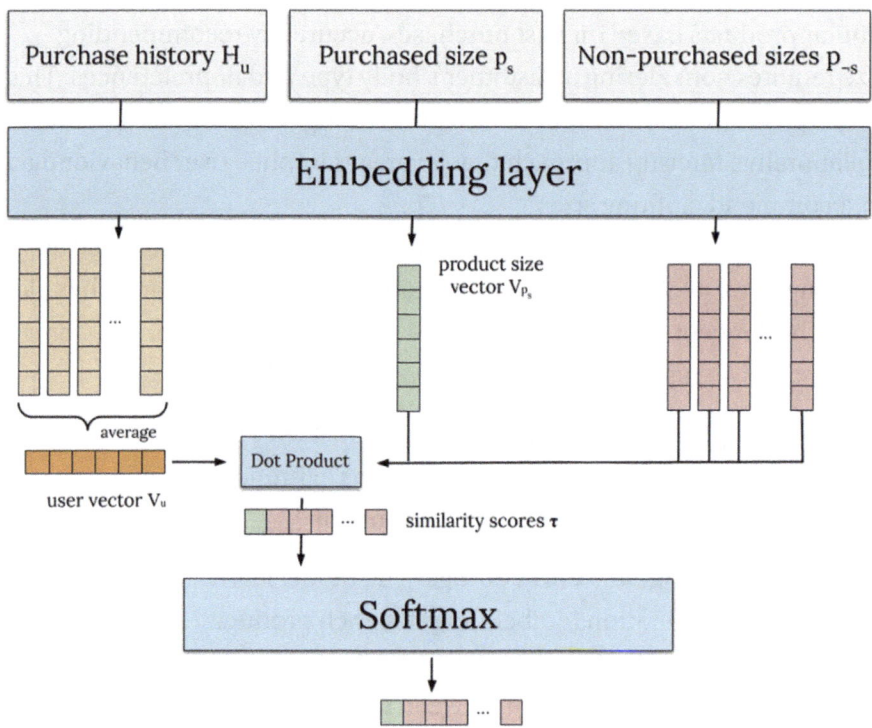

Figure 8-8. *Embedding enabled fit prediction*

Source: https://ceur-ws.org/Vol-2410/paper13.pdf

Benefits of PSE

- **Improved Size Recommendations:** By considering product-specific size information, PSE offers more accurate size recommendations compared to traditional methods.

- **Efficient Model Architecture:** The asymmetric framework reduces the number of parameters needed, making the model more efficient to train and deploy.

Reducing Customer Returns – Flagging Platform Abusers

In addition to helping the genuine customers with right size recommendations for reducing returns, identifying and taking actions against platform abusers are equally important to reduce the returns.

The Appriss Retail report found the returns of online purchases were worth $41 billion in total, where 35% (about $14 billion) was return frauds [14]. One report suggests that an astonishing 14% of returns have been found to be fraudulent [15].

Increased fraudulent returns warrants ecommerce platforms to develop defense against these practices. A strict return policy may lead to losing customers due to unpleasant return experiences; hence finding fraudulent customers and transactions are important to action against them only and proactively.

- **Delivery Charges on Return**: At one of the leading open marketplace platforms in India, based on ML model, it predicts whether a customer is likely to return the product, and the platform shows a message something like "If you returned, you will be charged shipping fee"; this is more to deter customers who have intention to return the product.

- **Validate at Return Initiation**: When a customer is initiating a return journey, based on the predicted abuse, the platform makes it mandatory for the customer to upload product images. The computer vision model validates whether the product return is fake or genuine.

- **QC on Reverse Pick Up**: Quality checks (QC) during reverse pickup involve inspecting returned products before they are collected from buyers. This process adds extra effort and costs for the delivery executives and the ecommerce platform. Implementing QC for all returns is not always feasible due to these additional costs. Instead, a machine learning model can optimize this process. By analyzing historical data on incorrect and genuine returns, the model can incorporate features related to buyers, products, geography, and other factors. When a buyer initiates a return, the model assesses whether to enable the QC flag based on these features. If the flag is activated, the delivery partner performs a quality check before pickup; otherwise, the check is omitted.

Non-deliverable Orders and Return to Origin (RTO) Transactions

5–8% of ecommerce transactions are returned to sellers due to non-delivery. Mainly, there are two key contributors of non-delivery of the transactions: customer intent issue (customers do not have any intention to accept the delivery, platform abusers) and bad delivery addresses.

These faulty addresses are one of the major reasons behind RTO or Return to Origin losses for ecommerce businesses [16].

"The economic impact of bad addresses in India is significant: Emerging World's estimates from the top industries indicate that poor addresses cost India $10–14B annually or ~0.5% of the GDP" [15].

There are a number of ways to reduce the bad addresses on the ecommerce platform. One option to help customers selecting address is using "Google Map". There are two challenges with making it mandatory, first customers from tier 2 and tier 3 places in India may find it challenging and the second it may adversely impact return rate as they may select any place. In summary, the conversion rate may take a hit due to additional steps in the order journey and it can make it easy for platform abuser to select any address to place order as they anyway do not intend to receive.

Identifying risky shipping addresses can significantly reduce RTO percentage and improve deliverability. "Decrease chances of RTO/cancellation by up to 30%" [16] by implementing an address validation process.

At my earlier organization, an ML-based non-deliverable address identification process was set up for identifying gibberish and non-deliverable addresses, leading to significant reduction in RTO percentage along with reduction in wrong return. The transactions with bad addresses were identified in near real time and shared for review and actions. Considering focus on customer service, only 50% of these transactions were actioned – cancel transaction, block the customer, or accept only prepaid transactions. For the other 50%, the RTO percentage was 3X with significantly higher returns and fake returns.

Address is only one symptom, not all customers have wrong intentions; probably they are not aware of providing addresses in a correct way. Hence, customer history could play a significant impact factor in taking address. Customer RTO Model and Bad Delivery Address Model in conjunction helped in improving efficacy of identifying platform abusers and reducing RTO rate.

Cancellations

Order cancellation is another driver of in-efficiency for the ecommerce platform The higher percentage of cancellations leads to missed sell due to inventory hold out, payment gateway charges for prepaid orders, operational costs (Order processing, picking, and packaging), and logistic cost (if cancelled after hand over to courier partner). Cancellations also have impact on inventory information distortion, leading to increases in total system costs [19].

The cancellations create bad seller experience due to increased costs, order processing, and inventory hold off.

Based on experience, it was observed that 4–6% of the transactions were canceled by the customers, and over half of these transactions were cancelled within an hour of placing an order. Card Cancellations (within 1 hour) are one of the significant and incur payment charges – around 1.5% of the transaction value.

Figure 8-9 visualizes the prominent reasons for order cancellations and their relative contributions.

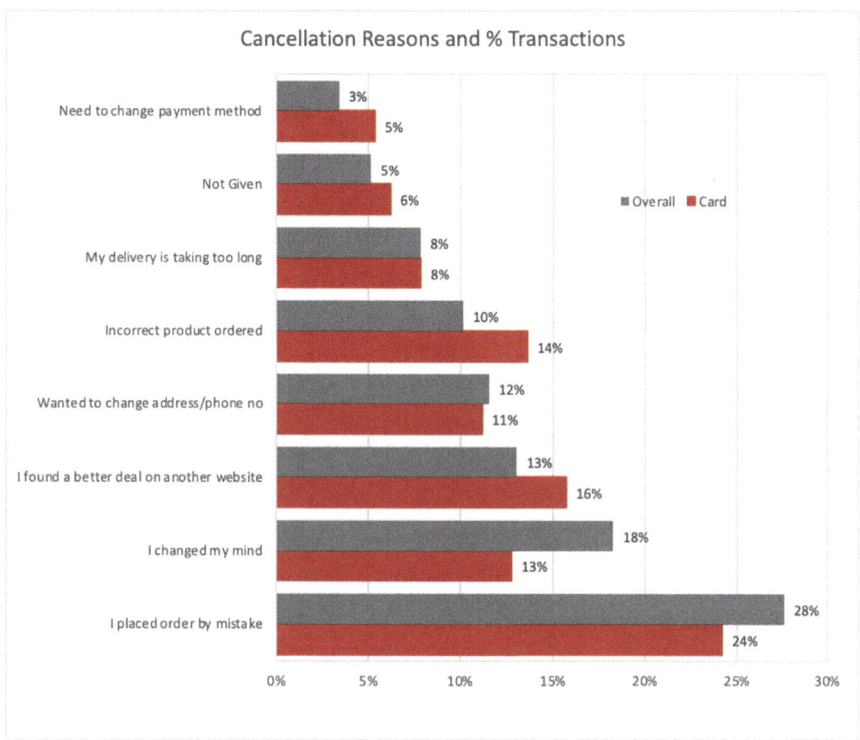

Figure 8-9. *Order cancellation reasons*

Longer delivery timelines or delayed delivery timelines along with product prices could be drivers of customers cancelling. But customers opt to select "I changed my mind" and "I placed an order by mistake."

ML Prediction for Improving Seller Experience and Cost: Once an order is confirmed on the ecommerce platform, the order is allocated to the seller and the seller operation team starts order processing. For the customers with higher chances of cancelling orders within an hour, the order can be put on hold for an hour before being assigned to the seller. This helps in reducing order processing effort and bad seller experience.

Summary

The total value of orders on an ecommerce platform is referred to as Gross Merchandising Value (GMV), while the final value of orders kept by buyers is known as Net Merchandising Value (NMV). The gap between GMV and NMV represents costs and adverse experiences for partners within the ecommerce ecosystem. Key drivers of this gap include cancellations, Return to Origin (RTO) orders, customer-initiated returns, and fake returns or damages.

The fulfillment journey involves multiple processes and partners, making technology, particularly machine learning and artificial intelligence (AI), a cost-effective and scalable solution for managing these challenges. Customer Initiated Returns (CIR), which occur after delivery, can stem from issues like product quality or fit. AI and ML can mitigate returns by offering personalized fit recommendations and virtual try-ons.

RTO, or non-delivery involves products returned to sellers due to issues such as incorrect addresses or buyer unavailability. Machine learning models help reduce RTO by proactively identifying and addressing problematic orders. Effective management of RTO is crucial for minimizing associated costs and enhancing ecommerce profitability.

Cancellations, whether initiated by customers or sellers, also pose significant challenges. AI and ML can assist in managing and reducing the costs associated with cancellations. Additionally, these issues often result in negative experiences for buyers and sellers, such as frustration from receiving poor-quality products or frequent cancellations. Abusive behaviors, like fraudulent returns or incorrect shipments, further exacerbate these problems. Machine learning models play a vital role in detecting and mitigating such abuses, improving the overall experience for all parties involved.

Leading ecommerce platforms are continuously leveraging and enhancing these technologies to improve buyer experience and operational efficiency. By addressing these challenges effectively, they aim to optimize processes and drive profitability in the competitive ecommerce landscape.

References

[1]   Maximizemarketresearch, India E-commerce Market: Rapid Digitalization and Government Initiatives Driving the Market Growth, `https://www.maximizemarketresearch.com/market-report/india-e-commerce-market/44404`

[2]   Rubal Sahni. Unlocking profit potential in retail returns, Apr-2023, `https://timesofindia.indiatimes.com/blogs/voices/207714/`

[3]   Rahi Bhattacharjee, eCommerce returns in India at 25-40%: What you should know for your D2C business, Mar-2022, `https://www.instamojo.com/blog/ecommerce-returns-in-india-what-you-should-know-for-your-d2c-business/`

[4]   Chargeback Gurus, The 12 Most Common Reasons for Returns, `https://www.chargebackgurus.com/blog/return-reasons`

[5]   Think With Google, Product Content Importance, `https://www.thinkwithgoogle.com/consumer-insights/consumer-trends/product-information-preference-statistics/`

[6]   Zoho Editor, Generate product descriptions with AI
        tools for your ecommerce store, Apr-2023, `https://`
        `www.zoho.com/commerce/insights/ecommerce-`
        `artificial-intelligence-ai-chatgpt-product-`
        `description.html`

[7]   Abdul-Saboor Sheikh, Romain Guigoures, others, A
        Deep Learning System for Predicting Size and Fit in
        Fashion E-Commerce, Jul-2019, `https://arxiv.org/`
        `pdf/1907.09844.pdf`

[8]   ALEXANDRE CANDEIAS, IVO SILVA, others,
        Size Recommendations for High-End Fashion
        Marketplaces, Jan-2024, `https://arxiv.org/`
        `pdf/2401.01978.pdf`

[9]   Kallirroi Dogani, Matteo Tomassetti, Sofie De
        Cnudde, Learning Embeddings for Product Size
        Recommendations, Jul-2019, `https://ceur-ws.org/`
        `Vol-2410/paper13.pdf`

[10]   Shreya Singh, G Mohammed Abdulla, Sumit Borar,
        Sagar Arora, Footwear Size Recommendation System,
        Jun-2018, `https://www.semanticscholar.org/reade`
        `r/3da5784751f908b9fa433170df26cb5f09c41c8a`

[11]   Vivek Sembium, Rajeev Rastogi, Atul Saroop, Srujana
        Merugu, Recommending product sizes to customers,
        Aug-2017, `https://cseweb.ucsd.edu/classes/fa17/`
        `cse291-b/reading/p243-sembium.pdf`

[12]   Rishabh Misra, Mengting, Julian McAuley,
        Decomposing Fit Semantics for Product Size
        Recommendation in Metric Spaces, Oct-2018,
        `https://cseweb.ucsd.edu/~jmcauley/pdfs/`
        `recsys18e.pdf`

[13]   Dong-Her Shih, Feng-Chuan Huang,Chia-Yi Chieh,Ming-Hung Shih,Ting-Wei Wu, Preventing Return Fraud in Reverse Logistics—A Case Study of ESPRES Solution by Ethereum, Jul-2021, `https://www.mdpi.com/0718-1876/16/6/121`

[14]   Lisa Jack, Regina Frei, Dr Sally-Ann Krzyzaniak, THE PROBLEMS & OPPORTUNITIES OF E-COMMERCE RETURNS, `https://www.ecrloss.com/research/buy-online-return-in-store`

[15]   Dr. Santanu Bhattacharya, Poor Addresses Cost India $10–14 Billion a Year, Dec-2018, `https://towardsdatascience.com/economic-impact-of-poor-addresses-in-india-10-14-billion-a-year-11cc97cb40fc`

[16]   Razorpay, Introducing Address Correction For E-Commerce Businesses, Dec-2020, `https://razorpay.com/blog/shipping-address-ecommerce/`

[17]   Pranav Mukul, E-commerce firms losing over 30 per cent of GMV due to cancellations/returns, Apr-2018, `https://indianexpress.com/article/business/business-others/e-commerce-firms-losing-over-30-per-cent-of-gmv-due-to-cancellations-returns-5125471/`

[18]   Digitalsunilsah, Flipkart Seller Order Cancellation Charges, Nov-2023, `https://ecomsprint.com/flipkart-seller-order-cancellation-charges/`

[19]   Hiroshi Ohta, Takayuki Furutani, EFFECT
OF CUSTOMER ORDER CANCELLATION
ON SUPPLY CHAIN INVENTORY, Feb-2010,
https://www.tandfonline.com/doi/
abs/10.1080/10170660409509385

[20]   Mckinsey, Returning to order: Improving returns
management for apparel companies, May-2021,
https://www.mckinsey.com/industries/retail/our-
insights/returning-to-order-improving-returns-
management-for-apparel-companies

Conclusion

Congratulations on completing a comprehensive exploration of ecommerce and its transformation through machine learning and artificial intelligence, particularly within the marketplace model.

In the initial chapters of this book, you delved into the intricacies of the ecommerce business model, examining the distinctions between marketplace and direct selling models. An essential aspect of understanding any ecommerce platform is grasping how it generates revenue.

Machine learning and artificial intelligence have a direct and measurable impact on a business's profit and loss statement. Key revenue streams in ecommerce include advertisements, commissions, and various fees. AI and ML play a crucial role in enhancing these revenue sources. For instance, advertising revenue, which monetizes active users on a marketplace, is largely driven by AI and ML algorithms that optimize ad targeting and placement. Similarly, AI and ML contribute to improving other revenue streams by enhancing efficiency, personalizing user experiences, and optimizing overall business operations.

Marketing, technology, and operations costs are key components of an ecommerce platform's profit and loss statement. Leveraging data and intelligence is crucial for achieving more with fewer resources. AI and ML play a significant role in optimizing these cost areas. For example, AI-driven marketing strategies can reduce customer acquisition costs (CAC) and enhance customer engagement through targeted CRM campaigns. Large Language Models (LLMs) facilitate the development and testing

© Ramgopal Prajapat 2024
R. Prajapat, *AI-Powered Ecommerce*, https://doi.org/10.1007/979-8-8688-0923-1

of systems, forecasting resource needs, and implementing preventive measures, all of which help to lower technology costs and mitigate potential challenges.

Operational efficiency is critical for delivering exceptional buyer experiences, but it often presents challenges. AI and ML-powered solutions can provide predictive and preventive measures to streamline operations and improve efficiency.

In Chapter 2 of the book, the focus is on the core ecommerce business model, particularly the marketplace platform. This platform connects buyers and sellers through a technological interface. The buy-side involves engaging buyers with features like search, recommendations, and personalization, all powered by AI and ML. On the sell-side, the platform supports seller onboarding, product creation, and functionality enhancements, many of which are augmented by AI. For instance, AI can aid in quality checks to reduce fake products, create product taxonomies, and generate product descriptions.

Can we list down at least 5 AI and ML use-cases that are discussed?

In a marketplace model, buyers are at the core of the business. The journey of engaging and converting buyers can be understood through three key funnels:

1. **Top Funnel**: This stage focuses on attracting buyers and prospective buyers to the platform. Effective strategies include targeted marketing campaigns, SEO, and other promotional activities to bring traffic to the platform.

2. **Mid Funnel**: Once buyers are on the platform, engagement becomes crucial. This involves enhancing their experience through personalized journeys, efficient search functionalities, relevant product displays, and personalized recommendations. The goal is to help buyers easily find products that match their interests and needs.

3. **Lower Funnel**: At this stage, the focus shifts to converting engagement into purchases. Streamlining the checkout process, offering data-driven coupons and promotions, and personalizing payment options are key strategies to improve the checkout experience and reduce cart abandonment. The aim is to maximize order value and secure successful transactions.

Having established an understanding of the ecommerce business model and marketplace technology, the next step is to delve deeper into critical components such as merchandising, search, recommendations, ranking, and personalization. These elements are essential for optimizing the buyer journey and enhancing overall platform performance.

Merchandising involves aligning target groups with the right products on an ecommerce platform. Buyers visit these platforms to purchase products, and it is crucial that they find these products easily and attractively. Category managers play a key role in ensuring that the platform features the right products and brands. Leveraging data and technology, they identify which brands and products are essential to buyers and ensure their availability. The site merchandising team then focuses on presenting these products in an engaging manner through homepage widgets and layouts. AI and ML technologies enhance this process by creating visually appealing presentations and defining widget sequences. Marketing teams utilize these visuals and promotional events to attract and engage buyers. This collaboration among category managers, site merchandising, and marketing, powered by AI and ML, ensures effective merchandising in ecommerce.

Search functionality is another critical component of an ecommerce platform. Although it appears as a simple search bar, it significantly impacts the business, influencing over 50% of the Gross Merchandising Value (GMV) with 3–4X higher conversion rates. To achieve this, the technology behind the search bar must effectively connect buyer queries with relevant products. Machine learning plays a vital role here, from classifying search queries and extracting entities to matching terms with product details. Advanced ranking models then ensure that the most relevant products are prominently displayed. This integration of machine learning optimizes search performance and enhances the overall user experience.

Semantic search and conversational AI are revolutionizing the search experience in ecommerce. Leading platforms like Amazon, Myntra, and Flipkart are each integrating generative AI functionalities to enhance buyer interactions. These advancements enable more intuitive and contextually relevant search results, improving the overall user experience.

Recommendations are a prominent use case of AI/ML in ecommerce. Amazon pioneered this approach in 1998 and has since evolved it into a versatile tool for enhancing buyer engagement. Recommendation algorithms now play a crucial role in influencing purchase decisions and shaping buyer journeys. Commonly used widgets powered by these algorithms include "Customers Who Bought This Also Bought," "Similar Products," "Recommended For You," "Inspired by Your Purchase History," and "Style Up," among others.

Collaborative Filtering, Content-Based Filtering, and their hybrid approaches are foundational methods for recommendation engines. These traditional techniques are complemented by advanced deep learning frameworks, such as Neural Collaborative Filtering, which enhance recommendation accuracy and personalization. By leveraging these sophisticated models, recommendation engines can provide more relevant and tailored product suggestions, significantly improving the buyer experience.

Both search results and recommendations are optimized through relevance ranking algorithms, which rearrange products to enhance their relevance for buyers. These algorithms utilize various performance, potential, and efficiency features to improve buyer experience and positively impact key business metrics, such as conversion rates.

Ranking algorithms extend beyond their roles in search and recommendation engines. They are also widely used as stand-alone solutions to achieve business objectives. For instance, they are instrumental in arranging widgets on the homepage and reordering categories within category widgets. Learning to Rank is a prominent application of these algorithms. The integration of deep learning frameworks with ranking algorithms is gaining significant traction in ecommerce.

What additional use cases for ranking algorithms have been discussed?

Personalization is a key differentiator across industries, with ecommerce leading the way in tailoring experiences for buyers. In the digital ecosystem of ecommerce, every touchpoint in the buyer journey can be personalized. Leading platforms like Myntra are setting the standard by creating highly personalized digital storefronts, including home pages and widget sequences tailored to individual buyers. Personalization extends to search results, recommendations, offers, promotions, and CRM campaigns, all driven by machine learning and AI technologies. Large Language Models further enhance personalization by crafting individualized tones for email marketing content. These initiatives not only foster customer loyalty but also create a competitive edge for ecommerce businesses. For instance, Stitch Fix has elevated personalization by integrating data science with human expertise to offer a personalized stylist experience for its users.

AI and ML are revolutionizing the buyer experience on marketplace platforms by enhancing search, recommendations, ranking, and personalization, which in turn boosts revenue and customer loyalty. However, the benefits of AI and ML extend beyond revenue growth. These technologies are also crucial for managing costs, mitigating risks, and improving operational efficiency.

In ecommerce, not all revenue captured as Gross Merchandise Value (GMV) translates into Net Merchandise Value (NMV) due to factors such as product returns, non-delivery (Return to Origin or RTO), cancellations by buyers and sellers, and fraudulent returns. Machine learning and AI algorithms play a vital role in addressing these issues and bridging the gap between GMV and NMV. For instance, AI can identify abusive customers who may contribute to RTOs, cancellations, or fake returns. Additionally, technologies like virtual try-ons and fit models help manage product returns and enhance the buyer experience, thereby improving overall efficiency.

Regardless of your role in the ecommerce business, you now have a wealth of practical use cases at your disposal to drive exceptional outcomes. The use cases discussed in this book are proven and have been successfully implemented across leading ecommerce platforms, including the author's own experiences. You can be confident that these approaches, when applied effectively in your context, will deliver impressive business results.

Now it's your turn to identify relevant use cases and begin implementing machine learning algorithms. I look forward to hearing about your successes and the impact these solutions have on your business.

Index

A

Add to cart rate, 56
AI, *see* Artificial intelligence (AI)
Amazon, 76, 112
Amazon India, 2, 104, 113, 152,
    194, 236
AOV, *see* Average order value (AOV)
AR, *see* Augmented Reality (AR)
Artificial intelligence (AI), 1, 2, 12,
    58, 65, 73, 105, 110, 143,
    174, 178, 184, 215, 216, 228,
    233, 238
ASP, *see* Average selling
    prices (ASP)
Augmented Reality (AR), 2, 32, 110
Average order value (AOV),
    49, 50, 70
Average selling prices (ASP), 56

B

Bandini Lehenga, 30, 31
Best Matching 25 (BM25), 92, 93
    See also Text matching
Bharat Interface for Money
    (BHIM), 34
BHIM, *see* Bharat Interface for
    Money (BHIM)

BI, *see* Business intelligence (BI)
BigBasket, 185
Bing ranking, 158
BLPs, *see* Brand listing
    pages (BLPs)
BM, *see* Brand marketing (BM)
BM25, *see* Best Matching
    25 (BM25)
Brand listing pages (BLPs), 53
Brand marketing (BM), 22
Brand prioritization, 68–73
Brand reputation, 67
B2C, *see* Business to
    consumer (B2C)
Business intelligence (BI), 49
Business to consumer (B2C), 3
Buyer, 77, 113–115, 117, 119, 177,
    178, 181, 186, 234–238
Buyer engagement, 32

C

CAC, *see* Customer acquisition
    cost (CAC)
Call-to-Actions (CTAs), 55
Call volume forecasting models, 35
Canceled order, 204
Cancellations, 226, 227

CAR, *see* Cart Addition Rate (CAR)

Cart Addition Rate (CAR), 55

Cash on delivery (COD), 34, 57, 195, 204

Category listing pages (CLPs), 53

CBF, *see* Content-based filtering (CBF)

CDPs, *see* Customer Data Platforms (CDPs)

CF, *see* Collaborative filtering (CF)

ChatGPT, 1, 104

Click-through rates (CTR), 9, 54, 65, 70, 95, 131, 152, 153, 174, 185, 190, 191

CIR, *see* Customer-initiated returns (CIR)

Cloud optimization, 13

CLPs, *see* Category listing pages (CLPs)

CM, *see* Contribution Margin (CM)

CM 1, *see* Contribution Margin 1 (CM 1)

CNNs, *see* Convolutional Neural Networks (CNNs)

COD, *see* Cash on delivery (COD)

Collaborative filtering (CF), 125–127, 236

Content-based filtering (CBF), 124, 141, 236

Contribution Margin (CM), 20

Contribution Margin 1 (CM 1), 22

Conversational search, 103, 104

Conversion rate (CVR), 56, 131

Convolutional Neural Networks (CNNs), 32, 144

Cosine similarity, 121

Cost per acquisition (CPA), 47

Cost-per-install (CPI), 50

Cost streams, 10, 11

CPA, *see* Cost per acquisition (CPA)

CPI, *see* Cost-per-install (CPI)

CRM, *see* Customer relationship management (CRM)

CTAs, *see* Call-to-Actions (CTAs)

Curated selection, 64

Customer acquisition cost (CAC), 50, 78

Customer cancellations, 206
order processing, 209
post-handover, 209
prepaid orders, 209

Customer Data Platforms (CDPs), 49, 179

Customer-initiated returns (CIR), 207, 208, 212–214, 228

Customer relationship management (CRM), 11, 22, 48, 51, 186, 188

Customer returns, 223, 224

CVR, *see* Conversion rate (CVR)

D

Data sparsity, 215

Deep learning, 42, 102, 103, 141, 142, 144, 186, 219

Delivery charges, return, 224

Deviation Analysis, 15

Digital marketing, 77

Digital marketplaces, 29
   buy side features, 37
   cost leakage, 33
   customer purchase journey, 36
   customer service team, 35
   functionalities, 37
   lower funnel, 56–58
   mid funnel, 51–56
   online transactions, 34
   sellers empowering, 38–44
   top funnel, 44–51
   TV and print media ads, 45
   See also E-commerce

Direct marketing (DM), 22

DM, *see* Direct marketing (DM)

E

E-commerce, 204, 233, 235, 237, 238
   and AI, 2
   AI/ML algorithms, 5
   algorithms, 1
   brand prioritization, 68–73
   business model, 3–6
   business value, 3
   buyers and sellers, 4
   category management, 66–68
   cost-effective way, 48
   economics, 20–23
   efficiency equation, 208–211
   high-level process, 83

Indian market, 2
   marketplaces, 30, 31
   merchandising, 64
   Myntra, 6, 7
   order fulfillment, 205–208
   personalization impact, 184–186
   price and promotion reviews, 71
   product category, 82
   recommendation engine, 110
   recommendation engines, 111–114
   search algorithms, 82
   search queries, 84–91
   search results, 95, 96
   search-savvy buyers, 82
   sellers empowering, 38–44
   semantic searches, 98–101
   strategy, 214
   technology and platform cost, 12–16
   website, 81

Euclidean distance, 121

F

Facebook, 47, 123, 159

Factorized models, 101

Fake returns, 208

Fashion-conscious buyers, 64

Fashion ecommerce
   algorithmic foundations, 135
   business model, 132
   challenges, 135
   cloud technology, 136

Fashion ecommerce (*cont.*)
  handcrafted features, 132
  similar buyers, 135
Fashion products (case study),
  166, 167
Filtering algorithms, 122–125
Fine-tuning, 102
Flipkart, 10, 104, 112, 162, 236
Forecasting business metrics, 15

G

Gaussian Mixture Model
  (GMM), 216
GenAI, *see* Generative Artificial
  Intelligence (GenAI)
Generative Artificial Intelligence
  (GenAI), 45, 46, 103, 104
GMM, *see* Gaussian Mixture
  Model (GMM)
GMV, *see* Gross Merchandise
  Value (GMV)
Google Map, 225
Google PageRank, 158–160
Gross Merchandise Value (GMV),
  8, 12, 20, 54, 74, 149, 177,
  198, 208, 210, 235, 238
GroupLens, 123

H

Home page personalization, 180–184
Hybrid recommendation
  algorithms, 128–130

I

IDF, *see* Inverse document
  frequency (IDF)
IKEA, furniture retailer, 33
Instagram, 47, 123
Interaction models, 101
Inverse document frequency (IDF),
  91, 93, 94
IP provider Television (IPTV), 114
IPTV, *see* IP provider
  Television (IPTV)
Item interaction matrix, 138

J

Jaccard similarity, 121

K

Key performance indicators
  (KPIs), 69
Kurta Set, 112

L

Large language models
  (LLMs), 13, 233
Latent Factor Models (LFMs), 101
Learning to Rank (LTR)
  techniques, 42, 52,
  167–169, 173
LFMs, *see* Latent Factor
  Models (LFMs)
LinkedIn, 123

LLMs, *see* Large language models (LLMs)

Location-based personalization, 179, 180

Long Short-Term Memory (LSTM) networks, 15

Lower funnel marketing, 235
 add-to-cart rate, 56
 CRM systems, 57
 ecommerce conversion, 57
 payment, 57
 prediction, 58

LSTM, *see* Long Short-Term Memory (LSTM) networks

LTR, *see* Learning to Rank (LTR) techniques

M

Machine learning (ML), 1, 12, 19, 52, 54, 57–59, 64–66, 73, 97, 114, 143, 150, 152, 174, 178, 184, 195, 233, 238
 ranking, 164–168
 search queries, 84–91
 size recommendations, 216–218
 user re-engagement case, 188, 189

Manhattan distance, 121

MAP, *see* Mean average precision (MAP)

Marketing team, 187–190

MAU, *see* Monthly active users (MAU)

Maya's Mystic Jewellery, 41, 42

Mean average precision (MAP), 168

Mean Reciprocal Rank (MRR), 167, 168

Meesho case study, 39, 40

Merchandising
 marketplace, 64–66

Mid funnel marketing, 234
 ecommerce platforms, 52
 product recommendations, 53
 site merchandising, 52
 site merchandising teams, 52
 visitors, 51

ML, *see* Machine learning (ML)

Monthly active users (MAU), 9

MRR, *see* Mean Reciprocal Rank (MRR)

Multi-category ecommerce, 185

Multi-stage issue management process, 15

Myntra, 2, 6, 7, 43, 64, 76, 96, 130, 138, 152, 163, 183, 184, 192, 193, 218, 236

N

Named Entity Recognition (NER), 86, 87

Natural Language Processing (NLP), 35, 36, 66, 86, 165, 184

NDCG, *see* Normalized Discounted Cumulative Gain (NDCG)

Netflix, 122, 123

NER, *see* Named Entity
     Recognition (NER)
Net Merchandise Value (NMV), 20,
     198, 208, 210, 238
Neural Collaborative Filtering
     (NCF), 140–142
NLP, *see* Natural Language
     Processing (NLP)
NMF, *see* Non-negative Matrix
     Factorization (NMF)
NMV, *see* Net Merchandise
     Value (NMV)
Non-deliverable orders, 225–227
Non-negative Matrix Factorization
     (NMF), 101
Normalized Discounted Cumulative
     Gain (NDCG), 169
Nykaa Fashion, 152

O

OMS, *see* Order management
     system (OMS)
One-size-fits-all strategy, 47
Operations cost, 16–20
Order confirmation, 204
Order management system (OMS),
     6, 41, 69

P

PageRank algorithm, 158–160
Payment sequence
     personalization, 196

Pay-per-click (PPC), 40
Paytm, 185
PDPs, *see* Product Detail/Detailed
     Pages (PDPs)
Pearson correlation coefficient, 121
Personalization, 237
     and contextual knowledge, 178
     ecommerce, 179, 184–186
     home page, 180–184
     location-based, 179, 180
     in marketing, 187–190
     recommendation engine, 177
     repeat users cases, 196
     search and
          recommendations, 178
     search engine, 190, 191
Personalized curations (case
     study), 74, 75
Personalized recommendations,
     115–118, 139–142
Personalized similar product
     recommendations,
     192, 193
Pinterest, 123
PLPs, *see* Product listing
     pages (PLPs)
Pondicherry, 109, 110
Popularity score, 162
Post order confirmation, 205, 206
PPC, *see* Pay-per-click (PPC)
Predictive maintenance, 13
Private-label products, 4
Product attributes, 165
Product content, 214, 215

Product Detail/Detailed Pages
(PDPs), 46, 51, 54, 56, 192
  recommendations, 136–139
Product listing pages (PLPs), 46, 51,
70, 75, 192
Product returns, 17, 34, 155
Product Size Embedding
(PSE) model
  asymmetric framework, 221
  benefits, 223, 224
  implicit signals, 221
  learning size embeddings, 221
  multi-class classification, 222
Product views to visits (PV/V), 70
PSE, *see* Product Size Embedding
(PSE) model
Purchase revenues, 131

Q

QC, *see* Quality checks (QC)
Quality checks (QC), 224
Query classification model, 86
Quora, 123

R

Ranking
  algorithms, 237
  deterministic model, 162, 163
  in Ecommerce, 150–156
  functional form, 157
  ML model, 164–168
  performance metrics, 155

popularity score, 162
product popularity, 154
products, 157, 159
recommendation engine, 149
recommendations, 169, 170
search functions, 160–162
similar product
  recommendations,
  171–174
Recommendation engines
  algorithms, 122–130
  Amazon, 114
  data, 119, 120
  e-commerce, 111–114
  evaluation, 130–132
  personalized
  recommendations,
  115–118
  personalized treatment, 114
  scenarios or functionalities,
  115, 116
  similarity measures, 120–122
Return on ad spend (ROAS), 47, 50
Return on investment (ROI), 49
Returns orders, 211, 212
Return to origin (RTO), 34, 207,
208, 211, 225–228
Revenue drivers
  AI/ML algorithms, 10
  commission, 8
  delivery fees, 8
  marketplace models, 7
RoBERTa model, 102

ROAS, *see* Return on ad
    spend (ROAS)
ROI, *see* Return on
    investment (ROI)
RTO, *see* Return to origin (RTO)

S

Science of similarity
    algorithms, 118
Search algorithms, 91, 92, 235
Search architecture, 96–99
Search embeddings, 102, 103
Search listing pages (SLPs), 90
Search personalizsation, 190, 191
    similar product
        recommendations,
        192–198
    Stitch Fix, 191, 192
Search term classification
    model, 84–86
Seller Portal, 7
Sellers, 78, 113, 238
    brands and, 71
    cancellations, 207, 209
    category team, 73
    empowering,
        e-commerce, 38–44
    selection case study, 41, 42
    statistics, 44
Sell side features, 43
Semantic search algorithms,
    99–102, 236
Sephora, 186

Similar products, 117
Singular Value Decomposition
    (SVD), 101
Size recommendations
    AI-based, 215, 216
    clothes sizes, 221
    ML model, 216–218
    Skip-gram, 218–221
Skip-gram, 218–221
SLPs, *see* Search listing
    pages (SLPs)
Smart order allocation, 17
Spotify, 123
Stitch Fix, 123, 191, 192
Style On, 112
Style Up, 142, 143
SVD, *see* Singular Value
    Decomposition (SVD)

T

Target groups (TGs), 63
Target variable, 166, 167
Tata CLiQ, 185, 194
Term frequency (TF), 93
Text matching
    IDF, 93, 94
    search algorithms, 91, 92
    TF, 93
TF, *see* Term frequency (TF)
TGs, *see* Target groups (TGs)
TikTok, 123
Tmall, 112
Top funnel marketing, 70, 234

business/category, 45
centralized strategy, 48
CRM, 51
data-driven approach, 47
data-driven personalization, 49
digital campaigns, 49
Ecommerce, 50
marketing teams, 50
multi-pronged approach, 48
SMS and WhatsApp, 49
social media campaigns, 47
social media platform, 46
TV and print media ads, 45
visual content creation,
    45, 46
Trending items, 116

U

UBCF, *see* User-Based Collaborative
    Filtering (UBCF)

Unified Payments Interface
    (UPI), 195
UPI, *see* Unified Payments
    Interface (UPI)
User-Based Collaborative
    Filtering (UBCF), 133

V

Virtual reality (VR), 32, 33
Visual content creation (case
    study), 45, 46
VR, *see* Virtual reality (VR)

W, X

Word2Vec Model, 218

Y, Z

YouTube, 47, 123

GPSR Compliance

The European Union's (EU) General Product Safety Regulation (GPSR) is a set of rules that requires consumer products to be safe and our obligations to ensure this.

If you have any concerns about our products, you can contact us on ProductSafety@springernature.com

In case Publisher is established outside the EU, the EU authorized representative is:

Springer Nature Customer Service Center GmbH
Europaplatz 3
69115 Heidelberg, Germany

Batch number: 08804053

Printed by Printforce, the Netherlands